Ma.al-MIU
L-tr
v.6

Algebra 1

LARSON
BOSWELL
KANOLD
STIFF

Applications • Equations • Graphs

Chapter 6
Resource Book

The Resource Book contains the wide variety
of blackline masters available for Chapter 6.
The blacklines are organized by lesson. Included
are support materials for the teacher as well as
practice, activities, applications, and assessment
resources.

McDougal Littell

A HOUGHTON MIFFLIN COMPANY

Evanston, Illinois • Boston • Dallas

Contributing Authors

The authors wish to thank the following individuals for their contributions to the Chapter 6 Resource Book.

Rita Browning
Linda E. Byrom
José Castro
Christine A. Hoover
Carolyn Huzinec
Karen Ostaffe
Jessica Pflueger
Barbara L. Power
Joanne Ricci
James G. Rutkowski
Michelle Strager

ISBN: 0-618-02044-6

3456789-CKI- 04 03 02 01

Contents

6 *Solving and Graphing Linear Inequalities*

Chapter Support		1-8
6.1	**Solving One-Step Linear Inequalities**	9-20
6.2	**Solving Multi-Step Linear Inequalities**	21-32
6.3	**Solving Compound Inequalities**	33-48
6.4	**Solving and Graphing Absolute Value Equations and Inequalities**	49-61
6.5	**Graphing Linear Inequalities in Two Variables**	62-75
6.6	**Stem-and-Leaf Plots and Mean, Median, and Mode**	76-90
6.7	**Box-and-Whisker Plots**	91-103
Review and Assess		104-117
Resource Book Answers		A1-A20

Contents

CHAPTER SUPPORT MATERIALS

Tips for New Teachers	p. 1	Prerequisite Skills Review	p. 5
Parent Guide for Student Success	p. 3	Strategies for Reading Mathematics	p. 7

LESSON MATERIALS

LESSON MATERIALS	6.1	6.2	6.3	6.4	6.5	6.6	6.7
Lesson Plans (Reg. & Block)	p. 9	p. 21	p. 33	p. 49	p. 62	p. 76	p. 91
Warm-Ups & Daily Quiz	p. 11	p. 23	p. 35	p. 51	p. 64	p. 78	p. 93
Alternative Lesson Openers	p. 12	p. 24	p. 36	p. 52	p. 65	p. 79	p. 94
Calc. Activities & Keystrokes			p. 37	p. 53	p. 66	p. 80	p. 95
Practice Level A	p. 13	p. 25	p. 39	p. 54	p. 67	p. 82	p. 96
Practice Level B	p. 14	p. 26	p. 40	p. 55	p. 68	p. 83	p. 97
Practice Level C	p. 15	p. 27	p. 41	p. 56	p. 69	p. 84	p. 98
Reteaching and Practice	p. 16	p. 28	p. 42	p. 57	p. 70	p. 85	p. 99
Catch-Up for Absent Students	p. 18	p. 30	p. 44	p. 59	p. 72	p. 87	p. 101
Coop. Learning Activities						p. 88	
Interdisciplinary Applications		p. 31			p. 73	p. 89	
Real-Life Applications	p. 19		p. 45	p. 60			p. 102
Math and History Applications			p. 46				
Challenge: Skills and Appl.	p. 20	p. 32	p. 47	p. 61	p. 74	p. 90	p. 103

REVIEW AND ASSESSMENT MATERIALS

Quizzes	p. 48, p. 75	Alternative Assessment & Math Journal	p. 112
Chapter Review Games and Activities	p. 104	Project with Rubric	p. 114
Chapter Test (3 Levels)	p. 105	Cumulative Review	p. 116
SAT/ACT Chapter Test	p. 111	Resource Book Answers	p. A1

Contents

Descriptions of Resources

This Chapter Resource Book is organized by lessons within the chapter in order to make your planning easier. The following materials are provided:

Tips for New Teachers These teaching notes provide both new and experienced teachers with useful teaching tips for each lesson, including tips about common errors and inclusion.

Parent Guide for Student Success This guide helps parents contribute to student success by providing an overview of the chapter along with questions and activities for parents and students to work on together.

Prerequisite Skills Review Worked-out examples are provided to review the prerequisite skills highlighted on the Study Guide page at the beginning of the chapter. Additional practice is included with each worked-out example.

Strategies for Reading Mathematics The first page teaches reading strategies to be applied to the current chapter and to later chapters. The second page is a visual glossary of key vocabulary.

Lesson Plans and Lesson Plans for Block Scheduling This planning template helps teachers select the materials they will use to teach each lesson from among the variety of materials available for the lesson. The block-scheduling version provides additional information about pacing.

Warm-Up Exercises and Daily Homework Quiz The warm-ups cover prerequisite skills that help prepare students for a given lesson. The quiz assesses students on the content of the previous lesson. (Transparencies also available)

Activity Support Masters These blackline masters make it easier for students to record their work on selected activities in the Student Edition.

Alternative Lesson Openers An engaging alternative for starting each lesson is provided from among these four types: *Application, Activity, Graphing Calculator,* or *Visual Approach.* (Color transparencies also available)

Graphing Calculator Activities with Keystrokes Keystrokes for four models of calculators are provided for each Technology Activity in the Student Edition, along with alternative Graphing Calculator Activities to begin selected lessons.

Practice A, B, and C These exercises offer additional practice for the material in each lesson, including application problems. There are three levels of practice for each lesson: A (basic), B (average), and C (advanced).

Contents

Reteaching with Practice These two pages provide additional instruction, worked-out examples, and practice exercises covering the key concepts and vocabulary in each lesson.

Quick Catch-Up for Absent Students This handy form makes it easy for teachers to let students who have been absent know what to do for homework and which activities or examples were covered in class.

Cooperative Learning Activities These enrichment activities apply the math taught in the lesson in an interesting way that lends itself to group work.

Interdisciplinary Applications/Real-Life Applications Students apply the mathematics covered in each lesson to solve an interesting interdisciplinary or real-life problem.

Math and History Applications This worksheet expands upon the Math and History feature in the Student Edition.

Challenge: Skills and Applications Teachers can use these exercises to enrich or extend each lesson.

Quizzes The quizzes can be used to assess student progress on two or three lessons.

Chapter Review Games and Activities This worksheet offers fun practice at the end of the chapter and provides an alternative way to review the chapter content in preparation for the Chapter Test.

Chapter Tests A, B, and C These are tests that cover the most important skills taught in the chapter. There are three levels of test: A (basic), B (average), and C (advanced).

SAT/ACT Chapter Test This test also covers the most important skills taught in the chapter, but questions are in multiple-choice and quantitative-comparison format. (See *Alternative Assessment* for multi-step problems.)

Alternative Assessment with Rubrics and Math Journal A journal exercise has students write about the mathematics in the chapter. A multi-step problem has students apply a variety of skills from the chapter and explain their reasoning. Solutions and a 4-point rubric are included.

Project with Rubric The project allows students to delve more deeply into a problem that applies the mathematics of the chapter. Teacher's notes and a 4-point rubric are included.

Cumulative Review These practice pages help students maintain skills from the current chapter and preceding chapters.

Tips for New Teachers

For use with Chapter 6

LESSON 6.1

TEACHING TIP You can help students understand how to graph an inequality. After students have solved for the variable, they should plot the number in the inequality on the number line. Then they can test a number to the left or to the right of the number plotted to determine whether it is a solution of the inequality. If it is, they should shade that side of the line. If not, they should shade the other side.

TEACHING TIP To emphasize that there can be more than one value that satisfies an inequality, ask your students to give you three numbers that would check in the original problem. They should do this after they have solved the inequality algebraically.

COMMON ERROR Students might reverse the inequality symbol whenever there is multiplication or division in the problem, or they might want to reverse the inequality symbol whenever there is a negative number involved in the problem. Remind your students that they should only reverse the inequality symbol if they have to multiply or divide by a negative number.

INCLUSION The expressions "less than" and "is less than" are especially tricky for students whose first language is not English. Make sure to show that the first expression is a subtraction but the second one is an inequality.

LESSON 6.2

TEACHING TIP Start the lesson by asking the students to solve some multi-step equations. You might want to review the need to "undo" the operations in the reverse order of the order of operations. Then, make the connection between solving an equation and solving an inequality and review the rules that students learned in Lesson 6.1.

INCLUSION Students might need to add some words denoting inequalities to their list of "key words." You can use this moment to review some of those words that they already know that translate to an inequality.

LESSON 6.3

INCLUSION To help students see the difference between compound inequalities with *and* or *or*, use some non-numerical examples. You might announce to your class that any student with a blue shirt *and* white socks will get an A today. How would that be different from announcing that students with a blue shirt *or* white socks will get an A?

TEACHING TIP The ideas of *union* and *intersection* underlie compound inequalities. Compound inequalities with *and* represent the *intersection* of two solution sets, whereas compound inequalities with *or* represent their *union*. You might want to discuss union and intersection with your students.

TEACHING TIP You can check students' understanding of compound inequalities by asking them to solve some unusual ones, such as $x > 1$ *and* $x > 5$, or $2x < 4$ *or* $x > -3$.

LESSON 6.4

COMMON ERROR Easy absolute value equations such as $|x| = 8$ might lead students to believe that the solutions for any absolute value equation are just the same number with opposite signs. For instance, students might say that the solutions for $|x - 5| = 2$ are ± 7, because they simply solve $x - 5 = 2$, and then write the answer as ± 7. Remind these students that they need to solve two separate equations to find the two solutions: $x - 5 = 2$ and $x - 5 = -2$.

TEACHING TIP An alternative method for solving absolute value inequalities is to first rewrite the inequality as an absolute value equation and find its two solutions. These two numbers will split the number line into three intervals. Now find what intervals belong to the solution. To do so, remember that absolute value inequalities such as $|ax + b| < c$ and $|ax + b| \leq c$ yield solutions in the interval between the two numbers, whereas those such as $|ax + b| > c$ and $|ax + b| \geq c$ result in solutions outside that same interval. Another option to determine what intervals are in the solution is to test numbers from each possible interval in the original inequality.

LESSON 6.5

COMMON ERROR Some students might get the erroneous idea that (0, 0) is the test point they should always use to decide which half-plane must be shaded. Remind them that the test point cannot be on the line separating the half-planes. You can use an example where the line passes through the origin to show students how to pick a test point.

TEACHING TIP Check for students' understanding of their graphs, especially when solving word problems such as Example 5 on page 362. In addition to graphing the solution, students should be able to list several different pairs of possible solutions. Notice that this problem requires all solutions to be integers, because the variables represent the number of gold and silver coins.

LESSON 6.6

INCLUSION "Stem-and-leaf" graphs and "bell-shape" curves make more sense when students know what *stem*, *leaf*, and *bell* mean. You might want to explain some words to students with limited English proficiency.

COMMON ERROR If a set of data includes a certain value more than once, remind students that they must include that value as many times as it occurs to calculate the mean and the median. Otherwise, some students might take only those values that are different from each other.

TEACHING TIP To check for students' understanding, have them calculate the mean, median, and mode of a small set of numbers (for example 2, 3, and 4). Then ask them whether the mean, median, and mode would change if a zero was added to the same set of numbers. Many students think that all three measures of central tendency stay the same because "adding a zero does not change anything."

TEACHING TIP You might want to ask your students to come up with examples of real-life data that would result in a bell-shape curve. Can they come up with any other real-life data that will *not* result in a bell-shape graph? What type of graph do they think is more common and why?

LESSON 6.7

INCLUSION Although most limited English students probably know what a box is, some of them might not know the meaning of whisker. "Box-and-whisker" plots will make more sense to them if you explain to them what a whisker is.

TEACHING TIP If your students get confused finding the quartiles, break down the process into smaller pieces where all they have to do is find medians. Be sure that they have rewritten the numbers in order. Start by asking them to find the median of all the numbers. Now, ask them to cover the upper half of the set of data and find the median again: they just found the first quartile. They can now cover the lower half of the data to find the median of the upper half: this is the third quartile.

Outside Resources

BOOKS/PERIODICALS

Kader, Gary D. "Means and MADS." *Mathematics Teaching in the Middle School* (March 1999); pp. 398–403.

Friel, Susan N. and William T. O'Connor. "Sticks to the Roof of Your Mouth?" *Mathematics Teaching in the Middle School* (March 1999); pp. 398–403.

SOFTWARE

Harvey, Wayne and Judah L. Schwartz. *The Function Supposer: Explorations in Algebra.* Newton, MA; Education Development Center, 1992.

VIDEOS

Algebra for Everyone: Videotape and Discussion Guide. Helpful ideas that use a broad range of approaches. Reston, VA; NCTM, 1991.

NAME _____ DATE _____

Parent Guide for Student Success

For use with Chapter 6

Chapter Overview One way that you can help your student succeed in Chapter 6 is by discussing the lesson goals in the chart below. When a lesson is completed, ask your student to interpret the lesson goals for you and to explain how the mathematics of the lesson relates to one of the key applications listed in the chart.

Lesson Title	Lesson Goals	Key Applications
6.1: Solving One-Step Linear Inequalities	Graph linear inequalities in one variable. Solve one-step linear inequalities.	• Running a Race • Steel Arch Bridge • Musical Instruments
6.2: Solving Multi-Step Linear Inequalities	Solve multi-step linear inequalities. Use linear inequalities to model and solve real-life problems.	• Population • Fly Fishing Business • Animated Films
6.3: Solving Compound Inequalities	Write, solve, and graph compound inequalities. Model a real-life situation with a compound inequality.	• Elevations • Prices of Fine Art • Solar System
6.4: Solving Absolute-Value Equations and Inequalities	Solve absolute-value equations and inequalities.	• Quality Control • Boxing • Fireworks
6.5: Graphing Linear Inequalities in Two Variables	Graph a linear inequality in two variables. Model a real-life situation using a linear inequality in two variables.	• Treasure Diving • Football • Nutrition
6.6: Stem-and-Leaf Plots and Mean, Median, and Mode	Make and use a stem-and-leaf plot to put data in order. Find the mean, median, and mode of data.	• Age Distribution • Manufacturing • Number of Households
6.7: Box-and-Whisker Plots	Draw a box-and-whisker plot to organize real-life data. Read and interpret a box-and-whisker plot of real-life data.	• Meteorology • Cellular Telephones • Skyscrapers

Study Strategy

Showing Your Work is the study strategy featured in Chapter 6 (see page 332). Encourage your student to record all steps on assignments since this helps to prevent errors and makes it easier to find and correct mistakes that do occur. When an answer is incorrect, you may be able to help your student in identifying the mistake and deciding how to correct the steps.

NAME _____ DATE _____

Parent Guide for Student Success

For use with Chapter 6

Key Ideas Your student can demonstrate understanding of key concepts by working through the following exercises with you.

Lesson	Exercise
6.1	Solve the inequality. $-5x < 45$ Is -10 a solution to the inequality? Is -8?
6.2	The Art Club is making crafts to raise money for art supplies. Each craft kit costs $3.84. The club spent $8 on advertising and plans to sell the crafts for $5 each. Write and solve an inequality to find how many crafts the club needs to sell to make a profit of at least $224.
6.3	Solve the inequality. $9 < 3 - 2x \le 11$
6.4	A box of cereal must weigh between 245 grams and 255 grams. Write an absolute-value inequality to represent this requirement.
6.5	You are buying party favors. Hats cost $0.22 each and small games cost $0.34 each. You have a maximum of $10 to spend on the favors. Write an inequality that models the different number of hats and small games you can purchase. Can you buy 20 of each? Can you buy 15 of each?
6.6	The numbers shown below represent the scores of ten students on the SAT math test. Find the mean, median, and mode of the data. 470, 512, 516, 485, 475, 523, 512, 518, 501, 497
6.7	Find the first, second, and third quartiles of the data. 19, 25, 16, 18, 17, 22, 22, 16, 24, 20

Home Involvement Activity

You will need: Paper, pencil, 10 coins

Directions: Toss up the ten coins and let them fall on a soft surface. Count the number of coins that land with heads up. Repeat for 25 trials, recording the data in a table. Find the mean, median, and mode of your data. Make a box-and-whisker plot.

Answers

6.1: $x > -9$; no; yes **6.2:** $5x - (3.84x + 8) \ge 224$; at least 200 **6.3:** $-4 \le x < -3$
6.4: $|x - 250| \le 5$ **6.5:** $0.22x + 0.34y \le 10$; no; yes **6.6:** mean 500.9, median 506.5, mode 512 **6.7:** 17, 19.5, 22

NAME _____ DATE _____

Prerequisite Skills Review

For use before Chapter 6

EXAMPLE 1 *Checking Possible Solutions*

Decide whether -2 is a solution of the inequality.

$5x^2 - 3x > 25$

SOLUTION

To check whether -2 is a possible solution, substitute -2 into the equation. If -2 makes the inequality true, then -2 is a solution.

$5(-2)^2 - 3(-2) > 25$

$5(4) + 6 > 25$

$20 + 6 > 25$

$26 > 25$ True

-2 is a solution

Exercises for Example 1

Decide whether -4 is a solution of the inequality.

1. $15x + x \geq 10$

2. $100 < 3(4.5x + 18)$

3. $\frac{1}{5}(25x + 60) = -48 - 4(x - 6)$

4. $-\frac{2}{9}(18x - 9) = -4\left(x - \frac{1}{2}\right)$

EXAMPLE 2 *Solving Multi-Step Equations*

Solve the equation.

$\frac{15}{2} + 23x = 42$

SOLUTION

$$\frac{15}{2} + 23x = 42 \qquad \text{Write original equation.}$$

$$\frac{15}{2} - \frac{15}{2} + 23x = 42 - \frac{15}{2} \qquad \text{Subtract } \frac{15}{2} \text{ from each side.}$$

$$23x = \frac{69}{2} \qquad \text{Simplify.}$$

$$x = \frac{69}{2} \cdot \frac{1}{23} \qquad \text{Divide both sides by 23.}$$

$$x = \frac{3}{2} \qquad \text{Simplify.}$$

The solution is $\frac{3}{2}$.

When you check the solution, substitute $\frac{3}{2}$ for x in the equation.

NAME _____ DATE _____

Prerequisite Skills Review

For use before Chapter 6

Exercises for Example 2

Solve the equation.

5. $7s + 2s = 90$

6. $9(x - 25x) = 81$

7. $6(y - 4) = 20.5$

8. $\frac{2}{9}x + \frac{1}{3} = \frac{19}{27}$

EXAMPLE 3 *Graphing an Equation*

Graph the equation.

$6x - y = 12$

SOLUTION

First write the equation in slope-intercept form, $y = mx + b$.

$6x - y = 12$	Write original equation.
$-6x + 6x - y = -6x + 12$	Subtract $6x$ from both sides of the equation.
$-y = -6x + 12$	Simplify.
$y = 6x - 12$	Multiply both sides of the equation by -1.

Now find the slope and the *y*-intercept.

$m = 6, \ b = -12$

Plot the point $(0, 6)$.

Draw a slope triangle to locate a second point on the triangle.

$m = \dfrac{6}{1} = \dfrac{rise}{run}$

Draw a line through the points.

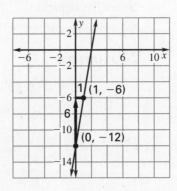

Exercises for Example 3

Graph the equation.

9. $5x + y = 8$

10. $-3x = 2y - 10$

11. $8x - y = 7 + 8x$

12. $x + 8y = 16$

CHAPTER

6

Chapter Support

NAME _____ DATE _____

Strategies for Reading Mathematics

For use with Chapter 6

Strategy: Reading Inequalities

Equations are easy to read. They have a simple meaning—this is equal to that. Inequalities may be more difficult to read because inequality symbols look similar but have different meanings. The graphs below show the different solutions that result from using other inequality symbols in the expression $x + 3 > -5$.

$x + 3 > -5$
$\quad x > -5 - 3$
$\quad\quad x > -8$

x is greater than -8.

$x + 3 \geq -5$
$\quad x \geq -5 - 3$
$\quad\quad x \geq -8$

x is greater than or equal to -8.

$x + 3 < -5$
$\quad x < -5 - 3$
$\quad\quad x < -8$

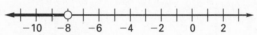

x is less than -8.

$x + 3 \leq -5$
$\quad x \leq -5 - 3$
$\quad\quad x \leq -8$

x is less than or equal to -8.

STUDY TIP

Reread Each Inequality

Mentally put on the brakes if you see an inequality sign. Stop and think about the meaning of the symbol. Then reread the whole inequality, substituting the word form of the symbol as you read.

STUDY TIP

Take Notes

If inequality symbols confuse you, then write each symbol with its meaning on the inside cover of your notebook for quick reference. Sketch a graph if that will help you remember.

Questions

1. Write the word form of each inequality.

   a. $x < 2y$ **b.** $x \geq 9$ **c.** $2x > -4$ **d.** $x - 8 \leq -1$

2. When you look at a graph of the solution of an inequality, what does a solid dot on the number line mean? What does an open dot on the number line mean?

3. Tell whether -2 is a solution of the inequality. Explain how you decided.

   a. $x > -4$ **b.** $x + 4 \leq 2$ **c.** $x - 4 < 2$ **d.** $x + 3 \geq 0$

4. What strategy do you use to help you remember the meaning of an inequality symbol when you are reading?

Visual Glossary

The Study Guide on page 332 lists the key vocabulary for Chapter 6 as well as review vocabulary from previous chapters. Use the page references on page 332 or the Glossary in the textbook to review key terms from prior chapters. Use the visual glossary below to help you understand some of the key vocabulary in Chapter 6. You may want to copy these diagrams into your notebook and refer to them as you complete the chapter.

GLOSSARY

graph of a linear inequality in one variable (p. 334) The set of points on a number line that represent all solutions of the inequality.

compound inequality (p. 346) Two inequalities connected by *and* or *or*.

solution of a linear inequality (p. 360) An ordered pair (x, y) is a solution of a linear inequality if the inequality is true when the values for x and y are substituted into the inequality.

half-plane (p. 360) In a coordinate plane, the region on either side of a boundary line.

Graphing Inequalities with One Variable

Graphing an inequality lets you see at a glance whether a given value is a solution of the inequality.

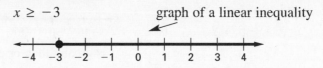

$x \geq -3$ graph of a linear inequality

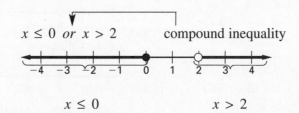

$x \leq 0 \;\; or \;\; x > 2$ compound inequality

$x \leq 0$ $x > 2$

The Graph of an Inequality in Two Variables

The graph of $x + y \geq 4$ is shown.

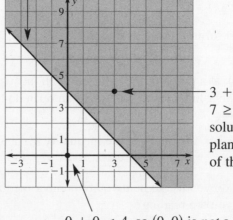

The line is the graph of the corresponding equation $x + y = 4$. The inequality symbol is $\leq$, so the line is solid.

$3 + 4 = 7$ and $7 \geq 4$, so $(3, 4)$ *is a* solution. This half-plane is the solution of the inequality.

$0 + 0 < 4$, so $(0, 0)$ is *not* a solution.

TEACHER'S NAME _____ CLASS _____ ROOM _____ DATE _____

Lesson Plan

1-day lesson (See *Pacing the Chapter*, TE pages 330C–330D)   **For use with pages 333–339**

GOALS   1. **Graph linear inequalities in one variable.**
          2. **Solve one-step linear inequalities.**

State/Local Objectives _____

_____

✓ Check the items you wish to use for this lesson.

STARTING OPTIONS
____ Prerequisite Skills Review: CRB pages 5–6
____ Strategies for Reading Mathematics: CRB pages 7–8
____ Warm-Up or Daily Homework Quiz: TE pages 334 and 321, CRB page 11, or Transparencies

TEACHING OPTIONS
____ Motivating the Lesson: TE page 335
____ Concept Activity: SE page 333
____ Lesson Opener (Application): CRB page 12 or Transparencies
____ Examples 1–4: SE pages 334–336
____ Extra Examples: TE pages 335–336 or Transparencies; Internet
____ Closure Question: TE page 336
____ Guided Practice Exercises: SE page 337

APPLY/HOMEWORK
Homework Assignment
____ Basic   22–54 even, 55–63, 66–68, 72, 76, 80, 86, 87
____ Average   22–54 even, 55–63, 66–68, 72, 76, 80, 86, 87
____ Advanced   22–54 even, 55–63, 66–70, 72, 76, 80, 86, 87

Reteaching the Lesson
____ Practice Masters: CRB pages 13–15 (Level A, Level B, Level C)
____ Reteaching with Practice: CRB pages 16–17 or Practice Workbook with Examples
____ Personal Student Tutor

Extending the Lesson
____ Applications (Real-Life): CRB page 19
____ Challenge: SE page 339; CRB page 20 or Internet

ASSESSMENT OPTIONS
____ Checkpoint Exercises: TE pages 335–336 or Transparencies
____ Daily Homework Quiz (6.1): TE page 339, CRB page 23, or Transparencies
____ Standardized Test Practice: SE page 339; TE page 339; STP Workbook; Transparencies

Notes _____

_____

_____

TEACHER'S NAME _____ CLASS _____ ROOM _____ DATE _____

Lesson Plan for Block Scheduling

Half-day lesson (See *Pacing the Chapter*, TE pages 330C–330D)   For use with pages 333–339

GOALS 1. **Graph linear inequalities in one variable.**
             2. **Solve one-step linear inequalities.**

State/Local Objectives _____

_____

_____

✓ **Check the items you wish to use for this lesson.**

STARTING OPTIONS
____ Prerequisite Skills Review: CRB pages 5–6
____ Strategies for Reading Mathematics: CRB pages 7–8
____ Warm-Up or Daily Homework Quiz: TE pages 334 and
         321, CRB page 11, or Transparencies

TEACHING OPTIONS
____ Motivating the Lesson: TE page 335
____ Concept Activity: SE page 333
____ Lesson Opener (Application): CRB page 12 or Transparencies
____ Examples 1–4: SE pages 334–336
____ Extra Examples: TE pages 335–336 or Transparencies; Internet
____ Closure Question: TE page 336
____ Guided Practice Exercises: SE page 337

APPLY/HOMEWORK
Homework Assignment
____ Block Schedule: 22–54 even, 55–63, 66–68, 72, 76, 80, 86, 87

Reteaching the Lesson
____ Practice Masters: CRB pages 13–15 (Level A, Level B, Level C)
____ Reteaching with Practice: CRB pages 16–17 or Practice Workbook with Examples
____ Personal Student Tutor

Extending the Lesson
____ Applications (Real-Life): CRB page 19
____ Challenge: SE page 339; CRB page 20 or Internet

ASSESSMENT OPTIONS
____ Checkpoint Exercises: TE pages 335–336 or Transparencies
____ Daily Homework Quiz (6.1): TE page 339, CRB page 23, or Transparencies
____ Standardized Test Practice: SE page 339; TE page 339; STP Workbook; Transparencies

CHAPTER PACING GUIDE	
Day	Lesson
1	Assess Ch. 5; **6.1 (all)**
2	6.2 (all); 6.3 (all)
3	6.4 (all)
4	6.5 (all)
5	6.6 (all)
6	6.7 (all)
7	Review/Assess Ch. 6

Notes _____

_____

_____

Lesson 6.1

NAME _____ DATE _____

WARM-UP EXERCISES

For use before Lesson 6.1, pages 333–339

Lesson 6.1

Solve each equation.

1. $5x - 7 = -12$

2. $6 = 3 - x$

3. $\dfrac{x}{4} = 12$

4. $8x = -32$

DAILY HOMEWORK QUIZ

For use after Lesson 5.7, pages 315–322

The table shows the number of U.S. high schools with computer networks from 1994 to 1997.

Years since 1994	Number of schools
0	6576
1	8159
2	9166
3	9565

1. Write a linear model for the number of high schools having computer networks.

2. Use the linear model to estimate the number of high schools with computer networks in 2002. Did you use linear interpolation or linear extrapolation?

Application Lesson Opener

For use with pages 334–339

1. Randall spent more at the mall than Karl. Karl spent $25. If a represents the amount Randall spent, which inequality describes the situation? Why?

   A. $a < 25$ **B.** $a \leq 25$

   C. $a > 25$ **D.** $a \geq 25$

2. Each day Grace studies with Trevor and then studies on her own. Trevor only studies with Grace. Grace studies a total of 1 hour each day. If t represents the time Trevor studies each day, which inequality describes this situation? Why?

   A. $t < 1$ **B.** $t \leq 1$

   C. $t > 1$ **D.** $t \geq 1$

3. Laura and Lyle collect coins. Laura has at least as many coins as Lyle. Lyle has 15 coins in his collection. If c represents the number of coins Laura has, which inequality describes the situation? Why?

   A. $c < 15$ **B.** $c \leq 15$

   C. $c > 15$ **D.** $c \geq 15$

4. Nina and Drew mow lawns. They work at most 10 hours each week. Drew always works 4 hours a week. If h represents the number of hours Nina works each week, which inequality describes the situation? Why?

   A. $h + 4 < 10$ **B.** $h + 4 \leq 10$

   C. $h + 4 > 10$ **D.** $h + 4 \geq 10$

NAME _____ DATE _____

Practice A

For use with pages 334–339

Write a verbal phrase that describes the inequality.

1. $x > 5$ 2. $x < -4$ 3. $3 \leq x$ 4. $-7 \geq x$

Write an inequality that describes the graph shown.

5.

6.

7.

8.

Sketch a graph of the inequality.

9. $x < 2$ 10. $x \geq 3$ 11. $4 < x$

12. $-1 \geq x$ 13. $x \geq 6$ 14. $x \leq 0$

Solve the inequality and graph its solution.

15. $x + 3 > 10$ 16. $x - 5 < 8$ 17. $4x \leq 16$ 18. $-2x \geq 2$

19. $x + 7 < -1$ 20. $x - 6 > -12$ 21. $\dfrac{x}{3} > 4$ 22. $\dfrac{x}{6} \leq -2$

23. $-5x \geq -15$ 24. $-8 < -4 + x$ 25. $18 > -9x$ 26. $14 \leq -\dfrac{x}{2}$

27. **Stereo**   You need at least $25 more to buy a stereo system. Write an inequality that describes how much money you need s. Graph the inequality.

28. **Siblings**   Maria is 15 years old. Let A represent the age of Maria's younger brother. Write an inequality for A. Graph the inequality.

29. **Temperature**   During a bitter cold week in January, the temperature in Seattle, Washington, did not exceed $-10°F$. Write an inequality that describes the temperature T in Seattle. Graph the inequality.

30. **Sales**   You want to buy a sweater that costs $42. The sweater might go on sale next week. Write an inequality that describes the price P of the sweater next week. Graph the inequality.

NAME _____ DATE _____

Practice B

For use with pages 334–339

Write an inequality that describes the graph shown.

1.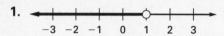
   -3 -2 -1 0 1 2 3

2. 0 1 2 3 4 5 6

3. -5 -4 -3 -2 -1 0 1

4. -6 -5 -4 -3 -2 -1 0

Sketch a graph of the inequality.

5. $x < -1$ 6. $x \geq 7$ 7. $6.5 < x$

8. $-1.8 \leq x$ 9. $x \geq 9$ 10. $x \leq -4$

Solve the inequality and graph its solution.

11. $x + 7 < 11$ 12. $x - 2 \leq 5$ 13. $3x > 6$ 14. $-2x \geq 8$

15. $x - 4 < -2$ 16. $-3 > x - 7$ 17. $\frac{x}{3} \geq -2$ 18. $\frac{x}{4} < 1.5$

19. $-8x \leq -24$ 20. $-7 < x + 2$ 21. $-6 \geq -\frac{x}{3}$ 22. $x + 3.5 > 8$

23. $-3.2x \geq 16$ 24. $-4.2 \leq x + 1.9$ 25. $2.3 > -\frac{x}{5}$

26. **Body Temperature**   Normal body temperature is 98.6°F. Write an inequality that describes the temperature T of people with above normal temperatures. Graph the inequality.

27. **Boiling Point**   Helium is the element that has the lowest boiling point, $-268.9°C$. Write an inequality that describes the boiling point b (in degrees Celsius) of any other element. Graph the inequality.

28. **Profit**   You make a profit of $5.25 from each magazine you sell. Write an inequality to show how many magazines m you need to sell to earn a minimum of $168. Graph the inequality.

29. **Basketball**   Tom has scored 181 points so far this basketball season. He needs to score 207 points to tie the school record for most points scored in a season. Let x represent the number of points Tom needs to score to tie or beat the record. Write an inequality for x. What is the least number of points Tom has to score? Graph the inequality.

30. **Elevations in California**   The lowest elevation in California is 282 feet below sea level. Let E represent the elevation of any location in California. Write an inequality for E. Graph the inequality.

Lesson 6.1

Practice C

For use with pages 334–339

Sketch a graph of the inequality.

1. $x > -3$

2. $x < 6$

3. $x \leq 0$

4. $x \geq 7.4$

5. $x < -2.3$

6. $-7.9 < x$

Solve the inequality and graph its solution.

7. $x - 9 > -11$

8. $x + 5 \leq 4$

9. $-6x \geq -24$

10. $9x \leq -81$

11. $-7 > x - 8$

12. $x + 10 < -23$

13. $\dfrac{x}{7} \geq 4$

14. $-\dfrac{x}{9} \leq -5$

15. $\dfrac{x}{2} \geq \dfrac{1}{8}$

16. $30 > -6x$

17. $x - 6.5 \geq 10$

18. $-4.7x \leq -15.04$

19. $-2.4x < 18$

20. $4.8 \leq -\dfrac{x}{3}$

21. $5.3 < x + 10.5$

22. *Largest Cat*   The world record for the heaviest domestic cat is a 46.95 pound tabby. Let C represent the weight of a domestic cat. Write an inequality for C. Graph the inequality.

23. *Viking Ships*   The longest Viking ship that has been found is not quite 95 feet in length. Let V represent the lengths of known Viking ships. Write an inequality for V. Graph the inequality.

24. *Work*   You earn $8.75 per hour at your job. Write an inequality to show how many full hours h you need to work to save at least $650. Graph the inequality.

25. *Melting Point*   Carbon is the element with the highest melting point, 3550°C. Write an inequality that describes the melting point m (in degrees Celsius) of any other element. Graph the inequality.

26. *Groceries*   At the grocery store, you need to spend $250 or more in two months to win a free turkey. A day before the contest ended, Mike had spent $213. Let x represent the amount of money Mike needs to spend to win the free turkey. Write an inequality for x. What is the least amount of money Mike has to spend? Graph the inequality.

27. *Marathon*   Paul ran a $26\frac{1}{5}$ mile marathon in $2\frac{3}{4}$ hours. Write an inequality to describe the average speeds s of runners who were faster than Paul. Graph the inequality.

NAME _____ DATE _____

Reteaching with Practice

For use with pages 334–339

GOAL **Graph linear inequalities in one variable and solve one-step linear inequalities**

> **VOCABULARY**
>
> The **graph** of a linear inequality in one variable is the set of points on a number line that represent all solutions of the inequality.
>
> **Equivalent inequalities** are inequalities that have the same solution(s).

EXAMPLE 1 *Graphing a Linear Inequality*

a. Graph the inequality $3 > x$.

b. Graph the inequality $x \geq 4$.

SOLUTION

a. Use an open dot for the inequality symbol < or >.

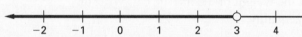

b. Use a closed dot for the inequality symbol ≤ or ≥.

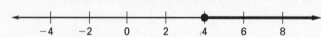

Exercises for Example 1

Graph the inequality.

1. $x \leq -1$ **2.** $x \geq 0$ **3.** $x < 0$

EXAMPLE 2 *Using Addition or Subtraction to Solve an Inequality*

Solve $x - 7 > -6$. Graph the solution.

SOLUTION

$$x - 7 > -6 \qquad \text{Write original inequality.}$$
$$x - 7 + 7 > -6 + 7 \qquad \text{Add 7 to each side.}$$
$$x > 1 \qquad \text{Simplify.}$$

The solution is all real numbers greater than 1. Check several numbers that are greater than 1 in the original inequality.

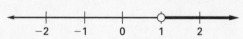

Algebra 1
Chapter 6 Resource Book

Reteaching with Practice

For use with pages 334–339

Lesson 6.1

Exercises for Example 2

Solve the inequality and graph its solution.

4. $1 > y - 1$ **5.** $x + 3 \le 0$ **6.** $k - 4 > -6$

EXAMPLE 3 *Using Multiplication or Division to Solve an Inequality*

a. Solve $-3x \ge -12$. **b.** Solve $\dfrac{n}{-2} < 5$. **c.** Solve $4y \le -8$.

SOLUTION

a. $-3x \ge -12$ Write original inequality.

$\dfrac{-3x}{-3} \le \dfrac{-12}{-3}$ Divide each side by -3 and reverse inequality symbol.

$x \le 4$ Simplify.

The solution is all real numbers less than or equal to 4. Check several numbers that are less than or equal to 4 in the original inequality.

b. $\dfrac{n}{-2} < 5$ Write original inequality.

$-2 \cdot \dfrac{n}{-2} > -2 \cdot 5$ Multiply each side by -2 and reverse inequality symbol.

$n > -10$ Simplify.

The solution is all real numbers greater than -10. Check several numbers that are greater than -10 in the original inequality.

c. $4y \le -8$ Write original inequality.

$\dfrac{4y}{4} \le \dfrac{-8}{4}$ Divide each side by positive 4.

$y \le -2$ Simplify.

The solution is all real numbers less than or equal to -2. Check several numbers that are less than or equal to -2 in the original inequality.

Exercises for Example 3

Solve the inequality and graph its solution.

7. $\dfrac{x}{4} < -1$ **8.** $-2a \ge -6$ **9.** $\dfrac{t}{-2} > 3$

NAME _____ DATE _____

Quick Catch-Up for Absent Students

For use with pages 333–339

The items checked below were covered in class on (date missed) _____

Activity 6.1: Investigating Inequalities (p. 333)

____ **Goal:** Determine when an operation changes an inequality.

Lesson 6.1: Solving One-Step Linear Inequalities

____ **Goal 1:** Graph linear inequalities in one variable. (p. 334)

Material Covered:

____ Student Help: Look Back

____ Example 1: Write and Graph a Linear Inequality

Vocabulary:

graph of a linear inequality, p. 334

____ **Goal 2:** Solve one-step linear inequalities. (pp. 335–336)

Material Covered:

____ Student Help: Look Back

____ Example 2: Using Subtraction to Solve an Inequality

____ Student Help: Study Tip

____ Example 3: Using Addition to Solve an Inequality

____ Student Help: Look Back

____ Example 4: Using Multiplication or Division to Solve an Inequality

Vocabulary:

equivalent inequalities, p. 335

____ Other (specify) _____

_____

Homework and Additional Learning Support

____ Textbook (specify) pp. 337–339 _____

_____

____ Internet: Extra Examples at www.mcdougallittell.com

____ *Reteaching with Practice* worksheet (specify exercises) _____

____ *Personal Student Tutor* for Lesson 6.1

NAME _____ DATE _____

Real Life Application:
When Will I Ever Use This?

For use with pages 334–339

Golf

Fifteen hundred years ago, Romans enjoyed playing *paganica*, a game in which a feather-filled ball was hit with a stick. Today, millions of people throughout the world enjoy a modern version of *paganica*—golf.

As do many other sports, golf has its own vocabulary. *Par* is the ideal number of strokes needed to complete a hole (or a course). One stroke less than par for a hole is called a *birdie*. One stroke more than par for a hole is called a *bogey*. An *eagle* is a score two less than par for a hole.

A player's *handicap* is essentially the average number of strokes he or she is above or below par for a course.

In Exercises 1 and 2, use the following information.

To qualify for the high school golf team, members must have a handicap of 20 or less.

1. Let h represent a high school golf team member's handicap. Write an inequality for h.

2. Graph the inequality.

In Exercises 3 and 4, use the following information.

During a golf tournament, the leader was hitting an average of 90 strokes or less.

3. Let s represent the leader's average score. Write an inequality for s.

4. Graph the inequality.

In Exercises 5 and 6, use the following information.

In order to stay on the golf team, members agree to spend at least 12 hours practicing their shots.

5. Let h represent the number of hours team members spend practicing their shots. Write an inequality for h.

6. Graph the inequality.

NAME _____ DATE _____

Challenge: Skills and Applications

For use with pages 334–339

1. If $a > b$ and $c < 0$, is $ac > bc$ or is $ac < bc$?

In Exercises 2–5, write an inequality.

2. x is at least -5

3. x is not more than 8

4. x is not less than 3

5. x is at most 2

In Exercises 6–7, use the following number line.

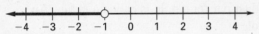

6. Write an inequality of the form $ax < b$ that has a solution matching the graph.

7. Write an inequality of the form $x + a < b$ that has a solution matching the graph.

In Exercises 8–9, use the following number line.

8. Write an inequality of the form $ax \leq b$ that has a solution matching the graph.

9. Write an inequality of the form $a - x \leq b$ that has a solution matching the graph.

In Exercises 10–15, *a* and *b* are real numbers such that $b > a > 0$. Tell whether the statement is *sometimes* true, *always* true, or *never* true. If it is sometimes true, give a set of values for which it is true and a set of values for which it is false.

10. $2b < a + b$

11. $\dfrac{a}{b} < \dfrac{b}{a}$

12. $b < b^2$

13. $a^2 < b^2$

14. $b^2 < a$

15. $-a^2 < -ab$

LESSON
6.2

Lesson Plan

1-day lesson (See *Pacing the Chapter,* TE pages 330C–330D)   **For use with pages 340–345**

GOALS   1. **Solve multi-step linear inequalities.**
2. **Use linear inequalities to model and solve real-life problems.**

State/Local Objectives _____

_____

✓ Check the items you wish to use for this lesson.

STARTING OPTIONS
____ Homework Check: TE page 337; Answer Transparencies
____ Warm-Up or Daily Homework Quiz: TE pages 340 and 339, CRB page 23, or Transparencies

TEACHING OPTIONS
____ Motivating the Lesson: TE page 341
____ Lesson Opener (Visual Approach): CRB page 24 or Transparencies
____ Examples 1–4: SE pages 340–342
____ Extra Examples: TE pages 341–342 or Transparencies; Internet
____ Closure Question: TE page 342
____ Guided Practice Exercises: SE page 343

APPLY/HOMEWORK
Homework Assignment
____ Basic   16–34 even, 37, 39, 43–51, 55, 60, 61, 64, 65
____ Average   16–34 even, 37, 39, 43–51, 55, 60, 61, 64, 65
____ Advanced   16–34 even, 37, 39, 43–53, 55, 60, 61, 64, 65

Reteaching the Lesson
____ Practice Masters: CRB pages 25–27 (Level A, Level B, Level C)
____ Reteaching with Practice: CRB pages 28–29 or Practice Workbook with Examples
____ Personal Student Tutor

Extending the Lesson
____ Applications (Interdisciplinary): CRB page 31
____ Challenge: SE page 345; CRB page 32 or Internet

ASSESSMENT OPTIONS
____ Checkpoint Exercises: TE pages 341–342 or Transparencies
____ Daily Homework Quiz (6.2): TE page 345, CRB page 35, or Transparencies
____ Standardized Test Practice: SE page 345; TE page 345; STP Workbook; Transparencies

Notes _____

_____

_____

Lesson 6.2

TEACHER'S NAME _____ CLASS _____ ROOM _____ DATE _____

Lesson Plan for Block Scheduling

Half-day lesson (See *Pacing the Chapter*, TE pages 330C–330D)   **For use with pages 340–345**

GOALS 1. **Solve multi-step linear inequalities.**
         2. **Use linear inequalities to model and solve real-life problems.**

State/Local Objectives _____

_____

_____

✓ **Check the items you wish to use for this lesson.**

STARTING OPTIONS

_____ Homework Check: TE page 337; Answer Transparencies

_____ Warm-Up or Daily Homework Quiz: TE pages 340 and
      339, CRB page 23, or Transparencies

TEACHING OPTIONS

_____ Motivating the Lesson: TE page 341

_____ Lesson Opener (Visual Approach): CRB page 24 or Transparencies

_____ Examples 1–4: SE pages 340–342

_____ Extra Examples: TE pages 341–342 or Transparencies; Internet

_____ Closure Question: TE page 342

_____ Guided Practice Exercises: SE page 343

APPLY/HOMEWORK

Homework Assignment (See also the assignment for Lesson 6.3.)

_____ Block Schedule: 16–34 even, 37, 39, 43–51, 55, 60, 61, 64, 65

Reteaching the Lesson

_____ Practice Masters: CRB pages 25–27 (Level A, Level B, Level C)

_____ Reteaching with Practice: CRB pages 28–29 or Practice Workbook with Examples

_____ Personal Student Tutor

Extending the Lesson

_____ Applications (Interdisciplinary): CRB page 31

_____ Challenge: SE page 345; CRB page 32 or Internet

ASSESSMENT OPTIONS

_____ Checkpoint Exercises: TE pages 341–342 or Transparencies

_____ Daily Homework Quiz (6.2): TE page 345, CRB page 35, or Transparencies

_____ Standardized Test Practice: SE page 345; TE page 345; STP Workbook; Transparencies

| CHAPTER PACING GUIDE ||
Day	Lesson
1	Assess Ch. 5; 6.1 (all)
2	**6.2 (all)**; 6.3 (all)
3	6.4 (all)
4	6.5 (all)
5	6.6 (all)
6	6.7 (all)
7	Review/Assess Ch. 6

Notes _____

_____

_____

Lesson 6.2

NAME _____ DATE _____

WARM-UP EXERCISES

For use before Lesson 6.2, pages 340–345

Solve the linear equation.

1. $2x - 1 = 7$

2. $5 = 3x - 7$

3. $2.5 = 3.3 - 4x$

4. $5 + \frac{2}{3}x = 1$

....

DAILY HOMEWORK QUIZ

For use after Lesson 6.1, pages 333–339

1. Graph $-1.5 < x$.

Solve the inequality and graph its solution.

2. $x - 4 \leq -1$

3. $2 > -1 - x$

Solve the inequality.

4. $1.8x \geq -8.1$

5. $-6 < -\dfrac{x}{7}$

Consider the inequality $2x + 3 \geq 5$.

1. Substitute the coordinate of the point graphed below into the inequality. Describe the result.

2. Use a red pencil to shade the number line to the right of the point. Choose several points that are in this shaded area. Substitute these values into the inequality. Describe the result.

3. Use a blue pencil to shade the number line to the left of the point. Choose several points that are in this shaded area. Substitute these values into the inequality. Describe the result.

4. Use your answers to Questions 1–3 to write the solution to the inequality.

Consider the inequality $5x - 6 \leq 9$.

5. Substitute the coordinate of the point graphed below into the inequality. Describe the result

6. Use a red pencil to shade the number line to the right of the point. Choose several points that are in this shaded area. Substitute these values into the inequality. Describe the result.

7. Use a blue pencil to shade the number line to the left of the point. Choose several points that are in this shaded area. Substitute these values into the inequality. Describe the result.

8. Use your answers to Questions 5–7 to write the solution to the inequality.

Lesson 6.2

NAME _____ DATE _____

Practice A

For use with pages 340–345

Solve the equation.

1. $2x - 7 = 5$

2. $-4 + 6x = 17$

3. $23 = 19 - x$

4. $5(x - 3) = 15$

5. $5x - 3 = 4x + 9$

6. $7 + 2x = -4x - 11$

Solve the inequality.

7. $x - 9 < -3$

8. $10 - x > 5$

9. $-3 \le 2x - 13$

10. $6x - 5 \ge 19$

11. $-2x + 9 > 15$

12. $-5x + 8 \le 3$

13. $4x + 6 > 14$

14. $-2x + 1 < 5x - 20$

15. $-x - 3 \ge -10x - 12$

16. $3x + 9 < 4x + 7$

17. $4x + 8 \le 2(x - 6)$

18. $-(x + 7) > 8x - 25$

School Enrollment **In Exercises 19–21, use the following information.**

In 1990 the enrollment at Trenton East High School was 620. From 1990 through 1996 the enrollment increased at an average rate of 30 students per year.

19. Write a linear model for the enrollment at Trenton East High School. Let x represent the number of years since 1990.

20. Trenton East was built to hold 800 students. Write a linear model for the maximum enrollment at Trenton East High School.

21. Write a linear inequality that represents the possible number of years since 1990 when the school's enrollment was less than the maximum capacity for which the school was built. Solve the inequality.

22. ***Amusement Park*** An amusement park charges $32 for admission and $6 to park your car. Write an inequality that represents the possible number of people that could go for $166. Solve the inequality. What is the maximum number of people that could go?

23. ***Long-Distance Services*** One long-distance company offers a plan such that each minute costs $.10 and each call has a $.10 service charge. Solve an inequality to find the maximum number of minutes you can talk for $3.00.

24. ***Chicken or Pork?*** From 1960 to 1990, the average annual U.S. consumption of pork dropped according to the linear model $y = -\frac{1}{3}x + 62$. The average annual consumption of chicken rose according to the linear model $y = \frac{7}{6}x + 31$. For which years did the U.S. consumption of chicken exceed the consumption of pork?

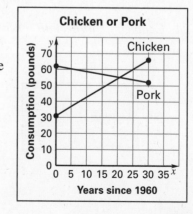

Chicken or Pork

Consumption (pounds)

Years since 1960

Practice B

For use with pages 340–345

Solve the inequality.

1. $7x - 30 < 19$

2. $-7 - 4x < 13$

3. $3x - 1 > 1$

4. $2x + 5 \leq 9$

5. $-3x - 5 < 16$

6. $-4x - 7 \geq -15$

7. $6x + 1 \leq -2$

8. $2x + 3 < 6x - 1$

9. $3x - 2 \geq 7x - 10$

10. $2x - 14 > 4x + 4$

11. $6x + 3 \leq 3(x + 2)$

12. $-2(x + 4) > 6x - 4$

13. $7 - 3x \geq x + 9$

14. $\frac{2}{3}x - 8 > -4$

15. $-12 \leq \frac{3}{5}x - 18$

16. $7x - 2 \leq -3(x - 2)$

17. $7 - 5x > 9 - 4x$

18. $3x - 4 \leq 2x - 4$

School Enrollment **In Exercises 19–20, use the following information.**

In 1990 the enrollment at Trenton East High School was 840. From 1990 through 1996 the enrollment increased at an average rate of 24 students per year.

19. Write a linear model for the enrollment at Trenton East High School. Let x represent the number of years since 1990.

20. Trenton East was built to hold 900 students. Write a linear inequality that represents the possible number of years since 1990 when the school's enrollment was less than the maximum capacity for which the school was built. Solve the inequality.

21. ***Water Park*** A water park charges $12 for admission and $5 to park your vehicle. Write an inequality that represents the possible number of people that could go for $50. Solve the inequality. What is the maximum number of people that could go?

22. ***Midway Games*** You want to go to the state fair and try your luck playing the games on the midway. The entrance fee is $5 and the games are each $1.50. Write an inequality that represents the possible number of games you can play if you have $25. Solve the inequality. What is the maximum number of games you can play?

23. ***Chicken or Pork?*** From 1960 to 1990, the average annual U.S. consumption of pork dropped from 62 pounds to 52 pounds. The average annual consumption of chicken rose from 31 pounds to 66 pounds. For which years did the U. S. consumption of chicken exceed the consumption of pork?

24. ***Geometry Connection*** Write an inequality for the values of x. Solve.

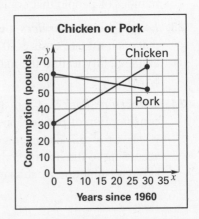

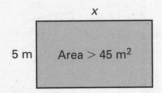

Lesson 6.2

Practice C

For use with pages 340–345

Solve the inequality.

1. $8x - 14 > 42$

2. $-9 - 6x < 21$

3. $7x - 5 \le 1$

4. $-4x + 8 \ge 20$

5. $-9x - 3 > -21$

6. $-2x - 13 < 17$

7. $6x + 5 \ge 7$

8. $x + 10 \le 3x - 8$

9. $15x - 33 > -2x + 18$

10. $5(x - 4) > -3x + 4$

11. $-2(7 - x) < -14 - 5x$

12. $3(2x + 9) \le 4(-x + 3)$

13. $-17x + 11 \ge 4x - 31$

14. $\frac{4}{3}x - 14 < 6$

15. $18 > -\frac{1}{6}x + 10$

16. $-5x + 21 \le -7x + 26$

17. $6x + 11 > 15 - 2x$

18. $-12x + 8 \ge 3(-9x + 1)$

19. *Whole Milk or Skim Milk?* From 1980 to 1996, the average annual U.S. consumption of skim milk rose from approximately 23 pounds to 55 pounds per person. The average annual consumption of whole milk dropped from approximately 140 pounds to 66 pounds per person. Which year did the U.S. consumption of skim milk first exceed the consumption of whole milk?

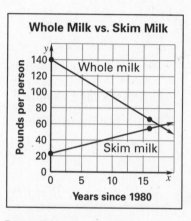

20. *Fitness Club* A membership to a fitness club has an initiation fee of $90. Each month you are billed an additional $30. Solve an inequality to find the maximum amount of time you can be a member for $600.

21. *Long-Distance Services* One long-distance telephone company offers a plan such that the first 20 minutes costs $.99 and each additional minute costs $.09. Solve an inequality to find the maximum amount of full minutes you can talk for $5.00.

22. *Geometry Connection* Write an inequality for the values of *x*. Solve.

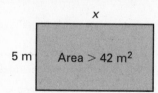

23. *Geometry Connection* Write an inequality for the values of *x*. Solve.

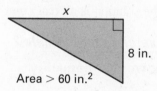

NAME _____ DATE _____

Reteaching with Practice

For use with pages 340–345

GOAL Solve multi-step linear inequalities and use linear inequalities to model and solve real-life problems

EXAMPLE 1 *Using More than One Step*

Solve $3n + 2 \leq 14$.

SOLUTION

$3n + 2 \leq 14$	Write original inequality.
$3n \leq 12$	Subtract 2 from each side.
$n \leq 4$	Divide each side by 3.

The solution is all real numbers less than or equal to 4.

Exercises for Example 1

Solve the inequality.

1. $5x - 7 > -2$ **2.** $9m + 2 \leq 20$ **3.** $13 + 4y \geq 9$

EXAMPLE 2 *Multiplying or Dividing by a Negative Number*

Solve $11 - 2x \geq 3x + 16$.

SOLUTION

$11 - 2x \geq 3x + 16$	Write original inequality.
$-2x \geq 3x + 5$	Subtract 11 from each side.
$-5x \geq 5$	Subtract $3x$ from each side.
$x \leq -1$	Divide each side by -5 and reverse inequality.

The solution is all real numbers less than or equal to -1.

Exercises for Example 2

Solve the inequality.

4. $8 > 5 - a$ **5.** $-4x + 2 \leq -22$ **6.** $-\dfrac{y}{2} + 3 \geq 0$

Reteaching with Practice

EXAMPLE 3 *Writing and Using a Linear Model*

You wash dishes in a restaurant and earn \$5.15 per hour. How many hours must you work to make at least \$200 to buy a new snowboard?

SOLUTION

Verbal Model

Hourly wage	·	Number of hours worked	≥	Desired income

Labels   Hourly wage = 5.15   (dollars per hour)

Number of hours worked = x (hours)

Desired income = 200   (dollars)

Algebraic Model

$5.15x > 200$ Write algebraic model.

$\dfrac{5.15x}{5.15} > \dfrac{200}{5.15}$ Divide each side by 5.15.

$x > 38.835\ldots$

You need to work at least 39 hours.

Exercises for Example 3

7. Rework Example 3 if you earn \$4.60 per hour.

8. Rework Example 3 if you need to make \$240 to buy a new snowboard.

NAME _____ DATE _____

Quick Catch-Up for Absent Students

For use with pages 340–345

The items checked below were covered in class on (date missed) _____

Lesson 6.2: Solving Multi-Step Linear Inequalities

____ **Goal 1:** Solve multi-step linear inequalities. (p. 340)

Material Covered:

    ____ Example 1: Using More than One Step

    ____ Example 2: Multiplying or Dividing by a Negative Number

____ **Goal 2:** Use linear inequalities to model and solve real-life problems. (pp. 341–342)

Material Covered:

    ____ Example 3: Writing a Linear Model

    ____ Activity: Investigating Problem Solving

    ____ Example 4: Writing and Using a Linear Model

____ Other (specify) _____

_____

Homework and Additional Learning Support

    ____ Textbook (specify) _pp. 343–345_ _____

_____

    ____ Internet: Extra Examples at www.mcdougallittell.com

    ____ *Reteaching with Practice* worksheet (specify exercises)_____

    ____ *Personal Student Tutor* for Lesson 6.2

Interdisciplinary Application

For use with pages 340–345

People in Flight

HISTORY In 1903 Orville and Wilbur Wright made the first flight in a self-propelled, heavier-than-air aircraft. Their first flight at Kitty Hawk, NC lasted only 12 seconds and 120 feet before returning to the ground. After continually improving upon their plane design, they came up with their first marketable plane in 1908. This plane, which was sold to the U.S. government, was capable of carrying two men for 125 miles at speeds up to 40 miles per hour.

1. In 1927 Charles Lindbergh became the first person to fly non-stop across the Atlantic Ocean. He traveled 3614 miles in $33\frac{1}{2}$ hours. The speed of planes increased each year between 1908 and 1927 approximately 3.5 miles per hour over the Wright brothers' plane's speed of 40 miles per hour. Write an inequality that describes the increase in flight speeds between 1908 and 1927.

2. In 1947 Charles "Chuck" Yeager broke the sound barrier in his Bell X-1 rocket plane. He attained speeds over 700 miles per hour. This is a speed increase of 29.6 miles per hour per year over Charles Lindbergh's 1927 flight. His flight speed was 108 miles per hour. Write an inequality representing the change from 1927 to 1947.

3. The average speed increase from the Wright brothers' flight to Yeager's flight was 16.9 miles per hour per year. Write an inequality representing the change from 1908 to 1947.

NAME _____ DATE _____

Challenge: Skills and Applications

For use with pages 340–345

In Exercises 1–6, solve the inequality.

1. $3x - \frac{1}{2} < \frac{1}{2}x + 7$

2. $4(5 - a) \geq 2(10 + 2a)$

3. $\frac{3}{4}w - 6 > 3 + w$

4. $7(2t + 1) \leq 2(7t - 3)$

5. $0.4(y + 9) < -1.2(8 - y)$

6. $4r - 7 \geq 13 - 4(5 - r)$

In Exercises 7–8, use the following information.

Music Club A offers each new subscriber 3 free tapes, after which each tape costs $7. Music Club B offers new subscribers 5 free tapes, after which each tape costs $9.

7. Write an inequality that states that the cost of n tapes from Club A is less than the cost of n tapes from Club B.

8. For what number of tapes is it cheaper to buy from Club A? Assume that a subscriber to either club wants at least 6 tapes.

In Exercises 9–10, write an inequality with a variable on both sides that satisfies the three given conditions.

9. • At least three operations are used.

• The inequality sign shown is $\geq$.

• The solution is $x \geq 2\frac{1}{2}$.

10. • At least one set of parentheses is used.

• The inequality sign shown is $<$.

• The solution is $x > -5$.

Algebra 1
Chapter 6 Resource Book

TEACHER'S NAME _____ CLASS _____ ROOM _____ DATE _____

Lesson Plan

1-day lesson (See *Pacing the Chapter,* **TE pages 330C–330D)**   **For use with pages 346–352**

GOALS 1. **Write, solve, and graph compound inequalities.**
2. **Model a real-life situation with a compound inequality.**

State/Local Objectives _____

_____

✓ Check the items you wish to use for this lesson.

STARTING OPTIONS
_____ Homework Check: TE page 343; Answer Transparencies
_____ Warm-Up or Daily Homework Quiz: TE pages 346 and 345, CRB page 35, or Transparencies

TEACHING OPTIONS
_____ Motivating the Lesson: TE page 347
_____ Lesson Opener (Activity): CRB page 36 or Transparencies
_____ Graphing Calculator Activity with Keystrokes: CRB pages 37–38
_____ Examples 1–6: SE pages 346–348
_____ Extra Examples: TE pages 347–348 or Transparencies; Internet
_____ Closure Question: TE page 348
_____ Guided Practice Exercises: SE page 349

APPLY/HOMEWORK
Homework Assignment
_____ Basic   12–38 even, 39, 40, 45, 50, 62, 65; Quiz 1: 1–15
_____ Average   12–38 even, 39, 40, 45, 50, 62, 65; Quiz 1: 1–15
_____ Advanced   12–38 even, 39, 40, 45, 46–48, 50, 62, 65; Quiz 1: 1–15

Reteaching the Lesson
_____ Practice Masters: CRB pages 39–41 (Level A, Level B, Level C)
_____ Reteaching with Practice: CRB pages 42–43 or Practice Workbook with Examples
_____ Personal Student Tutor

Extending the Lesson
_____ Applications (Real-life): CRB page 45
_____ Math & History: SE page 352; CRB page 46; Internet
_____ Challenge: SE page 351; CRB page 47 or Internet

ASSESSMENT OPTIONS
_____ Checkpoint Exercises: TE pages 347–348 or Transparencies
_____ Daily Homework Quiz (6.3): TE page 351, CRB page 51, or Transparencies
_____ Standardized Test Practice: SE page 351; TE page 351; STP Workbook; Transparencies
_____ Quiz (6.1–6.3): SE page 352; CRB page 48

Notes _____

_____

_____

Algebra 1
Chapter 6 Resource Book

Lesson 6.3

33

TEACHER'S NAME _____ CLASS _____ ROOM _____ DATE _____

Lesson Plan for Block Scheduling

Half-day lesson (See *Pacing the Chapter*, TE pages 330C–330D)   For use with pages 346–352

GOALS 1. **Write, solve, and graph compound inequalities.**
2. **Model a real-life situation with a compound inequality.**

State/Local Objectives _____

_____

_____

CHAPTER PACING GUIDE	
Day	**Lesson**
1	Assess Ch. 5; 6.1 (all)
2	6.2 (all); **6.3 (all)**
3	6.4 (all)
4	6.5 (all)
5	6.6 (all)
6	6.7 (all)
7	Review/Assess Ch. 6

✓ **Check the items you wish to use for this lesson.**

STARTING OPTIONS
____ Homework Check: TE page 343; Answer Transparencies
____ Warm-Up or Daily Homework Quiz: TE pages 346 and
        345, CRB page 35, or Transparencies

TEACHING OPTIONS
____ Motivating the Lesson: TE page 347
____ Lesson Opener (Activity): CRB page 36 or Transparencies
____ Graphing Calculator Activity with Keystrokes: CRB pages 37–38
____ Examples 1–6: SE pages 346–348
____ Extra Examples: TE pages 347–348 or Transparencies; Internet
____ Closure Question: TE page 348
____ Guided Practice Exercises: SE page 349

APPLY/HOMEWORK
Homework Assignment (See also the assignment for Lesson 6.2.)
____ Block Schedule: 12–38 even, 39, 40, 45, 50, 62, 65; Quiz 1: 1–15

Reteaching the Lesson
____ Practice Masters: CRB pages 39–41 (Level A, Level B, Level C)
____ Reteaching with Practice: CRB pages 42–43 or Practice Workbook with Examples
____ Personal Student Tutor

Extending the Lesson
____ Applications (Real-life): CRB page 45
____ Math & History: SE page 352; CRB page 46; Internet
____ Challenge: SE page 351; CRB page 47 or Internet

ASSESSMENT OPTIONS
____ Checkpoint Exercises: TE pages 347–348 or Transparencies
____ Daily Homework Quiz (6.3): TE page 351, CRB page 51, or Transparencies
____ Standardized Test Practice: SE page 351; TE page 351; STP Workbook; Transparencies
____ Quiz (6.1–6.3): SE page 352; CRB page 48

Notes _____

_____

_____

Lesson 6.3

NAME _____ DATE _____

WARM-UP EXERCISES

For use before Lesson 6.3, pages 346–352

Use the numbers -5, -4, -3, -2, -1, 0, 1, 2, 3, 4, 5.

1. Which numbers are less than or equal to -1 and greater than or equal to -2?

2. Which numbers are greater than 1 or less than -3?

3. Which numbers are less than or equal to -2 and less than or equal to 2?

4. Which numbers are greater than -1 or greater than 3?

···

DAILY HOMEWORK QUIZ

For use after Lesson 6.2, pages 340–345

Solve the inequality.

1. $4 - 3x \geq 7$

2. $2x - 3 < x + 6$

3. $-2x + 1 > 3(x - 8)$

4. A triangle has a height of 15 cm and a base of x cm. Write and solve an inequality for the values of x that give an area of at most 105 cm^2.

Activity Lesson Opener

For use with pages 346–352

SET UP: Work with a partner.
Choose the correct set of integers that solve the riddle.
Explain why your answer is correct.

1. We are integers that are less than 5 and greater than 1.
   What integers are we?

   A. 2, 3, 4   **B.** 1, 2, 3, 4

   C. 2, 3, 4, 5   **D.** all integers

2. We are integers that are greater than -2 and less than
   or equal to 3. What integers are we?

   A. all integers greater than -2

   B. all integers less than or equal to 3

   C. $-1, 0, 1, 2, 3$

   D. all integers

3. We are integers that are less than -3 or greater than 4.
   What integers are we?

   A. $-2, -1, 0, 1, 2, 3$

   B. $-3, -2, -1, 0, 1, 2, 3, 4$

   C. all integers

   D. all integers x such that $x < -3$ or $x > 4$

4. We are integers that are less than 2 or greater than 2. What integers are we?

   A. 0

   B. 2

   C. all integers except 2

   D. No integers fit this description.

Lesson 6.3

NAME _____ DATE _____

Graphing Calculator Activity

For use with pages 346–352

GOAL **To graph the solutions of two simple inequalities on the same number line**

In Lessons 6.1 and 6.2, you learned how to solve a simple inequality and graph its solution on a number line. You can use your graphing calculator to graph two simple inequalities on the same number line.

Activity

1 Solve the two simple inequalities below.

$x - 7 > -3$ $\qquad$ $-6x + 2 > 14$

2 Enter your solutions into Y_1 and Y_2 and plot the graphs of your solutions. (See keystrokes for special format.)

3 Do the two inequalities have any solutions in common?

4 Solve the two simple inequalities below.

$x + 2 < 5$ $\qquad$ $-4x - 1 < 19$

5 Repeat Steps 2 and 3 to graph the solutions of the inequalities from Step 5. (Be sure to clear out the solutions from Step 1.)

6 Why does the graph of both solutions turn out to be the entire number line?

7 Name three solutions that the inequalities have in common.

8 To graph only the solutions the inequalities have in common, enter both solutions in Y_1. (See keystrokes for special format.)

Exercises

1. Solve each pair of simple inequalities. Sketch both solutions on the same number line. Then check the graph with your graphing calculator.

a. $x + 5 < 4$ $\qquad$ **b.** $x + 5 < 5$ $\qquad$ **c.** $3x + 1 > -5$

$x - 7 > -6$ $\qquad$ $2x - 3 > 9$ $\qquad$ $-4x + 5 > -11$

2. Which pair of inequalities in Exercise 1 has solutions in common? Name two common solutions and then check your answer using your graphing calculator.

See page 38 for keystrokes.

LESSON
6.3
CONTINUED

NAME _____ DATE _____

Graphing Calculator Activity

For use with pages 346–352

TI-82

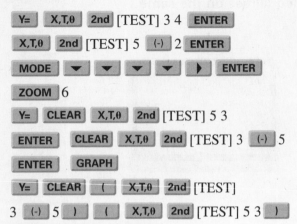

Y= | X,T,θ | 2nd | [TEST] 3 4 | ENTER
X,T,θ | 2nd | [TEST] 5 | (-) | 2 | ENTER
MODE | ▼ | ▼ | ▼ | ▼ | ▶ | ENTER
ZOOM | 6
Y= | CLEAR | X,T,θ | 2nd | [TEST] 5 3
ENTER | CLEAR | X,T,θ | 2nd | [TEST] 3 | (-) | 5
ENTER | GRAPH
Y= | CLEAR | (| X,T,θ | 2nd | [TEST]
3 | (-) | 5 |) | (| X,T,θ | 2nd | [TEST] 5 3 |)
ENTER | CLEAR | GRAPH

TI-83

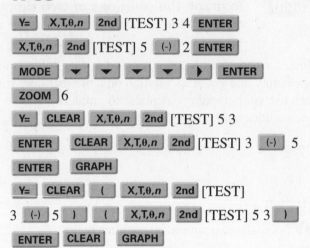

Y= | X,T,θ,n | 2nd | [TEST] 3 4 | ENTER
X,T,θ,n | 2nd | [TEST] 5 | (-) | 2 | ENTER
MODE | ▼ | ▼ | ▼ | ▼ | ▶ | ENTER
ZOOM | 6
Y= | CLEAR | X,T,θ,n | 2nd | [TEST] 5 3
ENTER | CLEAR | X,T,θ,n | 2nd | [TEST] 3 | (-) | 5
ENTER | GRAPH
Y= | CLEAR | (| X,T,θ,n | 2nd | [TEST]
3 | (-) | 5 |) | (| X,T,θ,n | 2nd | [TEST] 5 3 |)
ENTER | CLEAR | GRAPH

SHARP EL-9600c

Y= | X/θ/T/n | MATH | [F] 3 4
ENTER
X/θ/T/n | MATH | 5 | (-) | 2
ENTER
ZOOM | [A] 5
Y= | CL | X/θ/T/n | MATH | 5 3
ENTER
CL | X/θ/T/n | MATH | 3 | (-) | 5
ENTER
GRAPH
Y= | CL | (| X/θ/T/n | MATH | 3 | (-) | 5 |)
(| X/θ/T/n | MATH | 5 3 |) | ENTER
CL
GRAPH

CASIO CFX-9850GA PLUS

From the main menu, choose STAT.
Enter the following in List 1.

4 EXE 10 EXE

Enter the following in List 2.

(-) 10 EXE (-) 2 EXE

Enter the following in List 3.

1 EXE 1 EXE

SHIFT | F3 | F3 | EXIT | F1 | F6

Choose the following.

Graph Type: xyLine; XList: List1; YList: List3;
Frequency: 1; Mark Type: ■

EXIT | F6 | F2

Choose the following.

Graph Type: xyLine; XList: List2; YList: List 3;
Frequency: 1; Mark Type: ■

EXIT | F4 | F1 | ▼ | F1 | F6 | EXIT

Enter the following in List 1.

(-) 10 EXE 3 EXE

Enter the following in List 2.

(-) 5 EXE 10 EXE

F4 | F6 | EXIT

Enter the following in List 1.

(-) 5 EXE 3 EXE | F1

Lesson 6.3

NAME _____ DATE _____

Practice A

For use with pages 346–352

Write an inequality that represents the statement and graph the inequality.

1. x is less than 4 and greater than 0

2. x is greater than 4 or less than 2

3. x is at least 5 and at most 10

4. x is less than -3 or greater than 5

5. x is greater than -2 or less than -6

6. x is less than 4 and at least -3

Write an inequality that describes the graph shown.

7.
```
    +---•---+---+---+---•---+
    5   6   7   8   9  10  11
```

8.
```
    +---+---○---+---○---+---+
    5   6   7   8   9  10  11
```

9.
```
    +---+---○---+---+---•---+
   -4  -3  -2  -1   0   1   2
```

10.
```
    +---+---•---+---+---+---•---+
   -3  -2  -1   0   1   2   3
```

Sketch a graph of the inequality.

11. $-2 \leq x \leq 5$

12. $3 < x < 7$

13. $x \leq -3$ or $x > 1$

14. $x \leq 4$ or $x \geq 6$

15. $7 > x > 5$

16. $x > 5$ or $x < 0$

Solve the inequality and graph its solution.

17. $0 \leq x + 9 < 17$

18. $-14 < 7x < 21$

19. $x - 4 < -12$ or $2x \geq 12$

20. $-5 < -6 - x < 3$

21. $6 + 2x > 20$ or $8 + x \leq 0$

22. $-3x \leq 15$ or $5 + x < -11$

23. $-13 \leq 5 - 2x < 9$

24. $-22 > 11x$ or $4 + x > 4$

25. $2x + 1 > 13$ or $-18 > 7x + 3$

26. *Radar* A ship uses radar to detect approaching planes. A plane is shown as a blip on the radar. Use the diagram below to write an inequality that describes the distance of the plane from the ship.

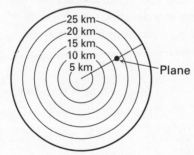

27. *Record High Temperature* The highest temperatures ever recorded in the United States for each month of the year are shown in the table. Write a compound inequality that represents the range of record high temperatures T in the United States.

Month	Temp.	Month	Temp.
Jan.	98°F	July	134°F
Feb.	105°F	Aug.	127°F
March	108°F	Sept.	126°F
April	118°F	Oct.	116°F
May	124°F	Nov.	105°F
June	129°F	Dec.	100°F

NAME _____ DATE _____

Practice B

For use with pages 346–352

Write an inequality that describes the graph shown.

1. ◄———+——+——○——+——+——●——+——►
   −4 −3 −2 −1 0 1 2

2. ◄——+——+——●——+——●——+——+——+——►
   6 7 8 9 10 11 12

3. ◄——+——○——+——+——+——○——+——►
   −5 −4 −3 −2 −1 0 1

4. ◄——+——●——+——+——+——○——+——►
   5 10 15 20 25 30 35

Sketch a graph of the inequality.

5. $-3 \leq x \leq 6$

6. $0 < x < 5$

7. $x < -1$ or $x \geq 2$

8. $x \leq 2$ or $x > 3$

9. $-2 > x \geq -10$

10. $x \geq 4$ or $x < -2$

Solve the inequality and graph its solution.

11. $3 < x - 3 \leq 5$

12. $-2 < 4 + x < 4$

13. $2x + 9 > 17$ or $5x + 10 < 10$

14. $3x - 6 \leq 0$ or $4x + 5 \geq -3$

15. $7 < 1 - x < 13$

16. $4 + 5x > 24$ or $16 + x \leq 17$

17. $1 \leq 3x \leq 6$

18. $-3 \leq 2x + 1 < 9$

19. $15 - 7x \leq -6$ or $6x - 3 < -6$

Solve the inequality and graph its solution. Then check graphically whether the given x-value is a solution by graphing the x-value on the same number line.

20. $-7 \leq 3 - 2x < 13; x = 1$

21. $-8 < 2x + 4 \leq -2; x = -3$

22. $4 - 3x \leq -8$ or $3x - 1 \leq 8; x = 0$

23. $5x + 1 < -14$ or $6x - 1 > 7; x = -3$

24. *Record Low Temperatures* The lowest temperatures ever recorded in the United States for each month of the year are shown in the table. Write a compound inequality that represents the range of record low temperatures T in the United States.

Month	Temp.	Month	Temp.
Jan.	$-70°F$	July	$10°F$
Feb.	$-69°F$	Aug.	$5°F$
March	$-50°F$	Sept.	$-9°F$
April	$-36°F$	Oct.	$-33°F$
May	$-15°F$	Nov.	$-53°F$
June	$2°F$	Dec.	$-59°F$

25. You live 4 miles from the convenience store and your friend lives 2 miles from the same store. (a) Find the minimum distance d between your home and your friend's home. (b) Find the maximum distance d between your home and your friend's home. (c) Write an inequality that describes the possible distances d between your home and your friend's home.

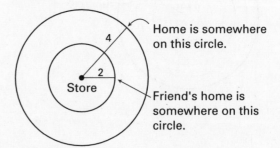

Home is somewhere on this circle.

Friend's home is somewhere on this circle.

Store

NAME _____ DATE _____

Practice C

For use with pages 346–352

Write an inequality that describes the graph shown.

1. ![number line from -12 to 0, closed dots at -8 and -4]

2. ![number line from -3 to 3, open circles at 0 and 2]

3. ![number line from 2.6 to 3.2, open circle at 2.8, closed dot at 2.9]

4. ![number line, closed dots at -5.0 and -4.8]

Sketch a graph of the inequality.

5. $-1 \leq x \leq 9$

6. $-8 < x < -3$

7. $x < 2 \text{ or } x \geq 4$

8. $x \leq \frac{2}{3} \text{ or } x > 5$

9. $-5.6 > x > -13.3$

10. $x > 0.1 \text{ or } x < 0.1$

Solve the inequality and graph its solution.

11. $-5 \leq -n - 6 \leq 0$

12. $-3 < 2x \leq 6$

13. $3x + 2 < -10 \text{ or } 2x - 4 > -4$

14. $-16 \leq 3x - 4 \leq 2$

15. $-8 < x - 11 < -6$

16. $x + 2 \leq 5 \text{ or } x - 4 \geq 2$

17. $-2 < -2n + 1 \leq 7$

18. $-7 < 6x - 1 < 5$

19. $-5x - 4 < -1.4 \text{ or } -2x + 1 > 11$

20. $-8 < \frac{2}{3}x - 4 < 10$

21. $-0.1 \leq 3.4x - 1.8 < 6.7$

22. $0.4x + 0.6 < 2.2 \text{ or } 0.6x > 3.6$

23. **Speed Limit** On Pennsylvania's interstate highway the speed limit is 65 mph. The minimum speed is 45 mph. Write a compound inequality that represents the speeds at which you may legally drive.

24. **Coupons** You have $60 and a coupon which allows you to take $10 off any purchase of $50 or more at a department store. Write an inequality that describes the possible retail value of the items you can buy if you use the coupon. Write an inequality that describes the different amounts of money you can spend if you use the coupon.

25. Consider the triangle below. Write a compound inequality that describes the possible lengths of the side of the triangle labeled x. Use the fact that the sum of the lengths of any two sides of a triangle is greater than the length of the third side.

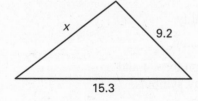

26. **Playing Tennis with a Friend** You live 5 miles from the tennis courts and 2 miles from your friend's house. Write an inequality that describes the distance between the tennis courts and your friend's house. Write an inequality that describes the distance you travel if you go to your friend's house and then to the tennis courts.

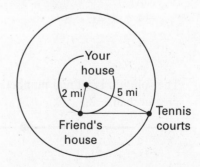

NAME _____ DATE _____

Reteaching with Practice

For use with pages 346–352

GOAL **Write, solve, and graph compound inequalities and model a real-life situation with a compound inequality**

> **VOCABULARY**
>
> A **compound inequality** consists of two inequalities connected by *and* or *or*.

EXAMPLE 1 **Writing Compound Inequalities**

a. Write an inequality that represents all real numbers that are less than 0 *or* greater than 3. Graph the inequality.

b. Write an inequality that represents all real numbers that are greater than or equal to -2 *and* less than 1. Graph the inequality.

SOLUTION

a. $x < 0$ *or* $x > 3$

b. $-2 \le x < 1$

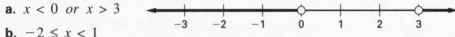

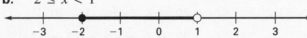

Exercises for Example 1

Write an inequality that represents the statement and graph the inequality.

1. x is greater than -4 *and* less than or equal to -2

2. x is greater than 3 *or* less than -1

EXAMPLE 2 **Solving a Compound Inequality with And**

Solve $-9 \le -4x - 5 < 3$. Graph the solution.

SOLUTION

Isolate the variable x between the two inequality symbols.

$-9 \le -4x - 5 < 3$	Write original inequality.
$-4 \le -4x < 8$	Add 5 to each expression.
$1 \ge x > -2$	Divide each expression by -4 and *reverse* both inequality symbols.

The solution is all real numbers that are less than or equal to 1 *and* greater than -2.

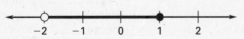

NAME _____ DATE _____

Reteaching with Practice

For use with pages 346–352

Exercises for Example 2

Solve the inequality and graph the solution.

3. $-3 < 2x + 1 \leq 7$ **4.** $-6 < -3 + x < -4$ **5.** $2 \leq -3x + 8 < 17$

EXAMPLE 3 ## *Solving a Compound Inequality with Or*

Solve $5x + 1 < -4$ *or* $6x - 2 \geq 10$. Graph the solution.

SOLUTION

You can solve each part separately.

$$5x + 1 < -4 \quad or \quad 6x - 2 \geq 10$$
$$5x < -5 \quad or \quad 6x \geq 12$$
$$x < -1 \quad or \quad x \geq 2$$

The solution is all real numbers that are less than -1 *or* greater than or equal to 2.

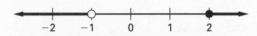

Exercises for Example 3

Solve the inequality and graph the solution.

6. $2x - 3 < 5$ *or* $3x + 1 \geq 16$ **7.** $-4x + 2 < 6$ *or* $2x \leq -6$

EXAMPLE 4 ## *Modeling with a Compound Inequality*

In 1985, a real estate property was sold for $145,000. The property was sold again in 1999 for $211,000. Write a compound inequality that represents the different values that the property was worth between 1985 and 1999.

SOLUTION

Use the variable v to represent the property value. Write a compound inequality with *and* to represent the different property values.

$$145,000 \leq v \leq 211,000$$

Exercise for Example 4

8. Rework Example 4 if the property was sold in 1985 for $172,000 and was sold again in 1999 for $226,000.

NAME _____ DATE _____

Quick Catch-Up for Absent Students

For use with pages 346–352

The items checked below were covered in class on (date missed) _____

Lesson 6.3: Solving Compound Inequalities

____ **Goal 1:** Write, solve, and graph compound inequalities. (pp. 346–347)

Material Covered:

____ Example 1: Writing Compound Inequalities

____ Example 2: Compound Inequalities in Real Life

____ Example 3: Solving a Compound Inequality with And

____ Example 4: Solving a Compound Inequality with Or

____ Student Help: Study Tip

____ Example 5: Reversing Both Inequality Symbols

Vocabulary:

compound inequality, p. 346

____ **Goal 2:** Model a real-life situation with a compound inequality. (p. 348)

Material Covered:

____ Example 6: Modeling with a Compound Inequality

____ Other (specify) _____

_____

Homework and Additional Learning Support

____ Textbook (specify) pp. 349–352 _____

_____

____ Internet: Extra Examples at www.mcdougallittell.com

____ *Reteaching with Practice* worksheet (specify exercises) _____

____ *Personal Student Tutor* for Lesson 6.3

NAME _____ DATE _____

Real Life Application:
When Will I Ever Use This?

For use with pages 346–352

The Value and Cost of Education

Your future income will be greatly influenced by the level of education you attain. According to 1994 income statistics published by the U.S. Bureau of the Census, the average annual earnings by level of education were as follows.

Level of Education	Average Annual Earnings
Did not finish high school	$12,809
High school diploma	$18,737
Associate's degree	$24,398
Bachelor's degree	$32,629
Master's degree	$40,368
Professional	$74,560

In Exercises 1–3, write a compound inequality that describes the situation.

1. An individual who has a Master's degree is earning more than average but less than a professional.

2. A high school graduate is earning at least the national average but less than someone with a Bachelor's degree.

3. A person with an Associate's degree is earning less than average but more than a high school graduate.

College Costs

In Exercises 4–5, use the following information.

If continuing your education is one of your goals, planning ahead and saving money is important. It is never too late to start saving. Education costs continue to increase each year and expenses range widely from one institution to another. When planning for education expenses, items such as books, room and board, transportation, and incidentals must also be taken into consideration. Four-year institutions range from $6,000 per year to $30,000 per year, while two-year business and trade schools range from $5,000 per year to $15,000 per year.

4. Write compound inequalities that describe each type of school.

5. Graph the inequalities from Exercise 4.

Math and History Application

For use with page 352

HISTORY   Communication has been an important part of life throughout history. The earliest written communications were in the form of pictures. Cave paintings dating to 30,000 B.C. have been found in France's Grotte Chauvet. These paintings represented animals and objects, and told stories. The earliest known form of writing is cuneiform, which was developed by the Sumerians. In cuneiform, symbols and pictures were used to represent words. The earliest forms of cuneiform were found on clay tablets dating to 3300 B.C. One of the earliest alphabets was developed about 1000 B.C. by the Phoenicians. Some of the letters in the English alphabet are from the Phoenician alphabet. Today there are about 6000 languages spoken around the world.

The ways in which messages are sent has become more sophisticated. Paper and the first printing blocks were invented in China around A.D. 100. The first printing press was made by Johannes Gutenberg around A.D. 1450. The printing press allowed the spread of ideas, facts, and knowledge to many people quicker and cheaper than before its invention.

The first postal systems used runners or couriers on horseback to carry messages. Cows, camels, and reindeer have been used in postal systems throughout the world. Written messages can now be sent by mail, fax, or e-mail. Spoken message can be sent by telephone, radio, television, and cellular telephone.

MATH   Write an inequality to represent the time it took to send a message.

1. The first message sent from London to New York by sailing ship took up to 9 weeks.

2. The first transatlantic telegraph was in 1858 and took less than 3 seconds.

3. Fast sailing clipper ships sailed across the Atlantic Ocean carrying mail in at least 12 days.

4. The first transcontinental airmail route took between 29 and 33 hours.

5. Arrange the facts listed below on a number line.

   a. the fax was first demonstrated in 1851

   b. the radio was invented in 1895

   c. the first television was demonstrated in 1926

   d. the first pony express delivery was in 1860

   e. the first telephone was patented in 1876

Challenge: Skills and Applications

For use with pages 346–352

1. If $a < x < b$ has no solution, what is the relationship between a and b?

2. If the compound inequality $x < a$ or $x > b$ has every real number as a solution, what is the relationship between a and b?

3. If $a < x < b$ and $c < 0$, write a true inequality for ac, cx, and bc.

4. Solve the inequality $-2 < 4x - 3\frac{1}{2} \leq 9$.

5. Graph the solution from Exercise 4.

6. Is $x = -1$ a solution to the inequality from Exercise 4?

7. Is $x = 2\frac{2}{3}$ a solution to the inequality from Exercise 4?

8. Solve the inequality $2.5x + 4 \leq 3.25$ or $3x - 7.3 > -4.6$.

9. Graph the solution from Exercise 8.

10. Is $x = -0.5$ a solution to the inequality from Exercise 8?

11. Is $x = 0.5$ a solution to the inequality from Exercise 8?

In Exercises 12–14, use the following information.

Two rivers combine to make another. The flow rate of the combined river is less than or equal to the sum of the flow rates of the two tributaries, but it is greater than or equal to either tributary's individual flow rate. One of the three rivers flows at a rate of 52 gallons per minute. Another flows at a rate of 38 gallons per minute.

12. Find the minimum flow rate f of the third river.

13. Find the maximum flow rate f of the third river.

14. Write an inequality that describes the possible flow rates f of the third river.

Lesson 6.3

NAME _____ DATE _____

Quiz 1

For use after Lessons 6.1–6.3

1. Solve $-11 + x \geq -3$ and graph the solution. *(Lesson 6.1)*

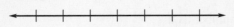

2. Solve $-\dfrac{x}{4} < 7$ and graph the solution. *(Lesson 6.1)*

3. The highest waterfall in the world is Angel Falls in Venezuela at 1000 meters. Write an inequality that describes the height h (in meters) of every other waterfall. Graph the inequality. *(Lesson 6.1)*

4. Solve $-4x - 3 > -5x + 8$. *(Lesson 6.2)*

5. Write an inequality for the value of x. *(Lesson 6.2)*

   Area ≤ 48 square centimeters

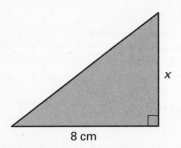

8 cm

6. Solve $-5 \leq 2x + 3 < 7$ and graph the solution. *(Lesson 6.3)*

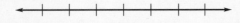

7. Solve $2x + 3 \leq 5$ or $3x - 4 > 8$ and graph the solution. *(Lesson 6.3)*

Answers

1. _____
   Use number line at left.

2. _____
   Use number line at left.

3. _____
   Use number line at left.

4. _____

5. _____

6. _____
   Use number line at left.

7. _____
   Use number line at left.

Review and Assess

TEACHER'S NAME _____ CLASS _____ ROOM _____ DATE _____

Lesson Plan

2-day lesson (See *Pacing the Chapter,* TE pages 330C–330D)   **For use with pages 353–359**

GOALS 1. **Solve absolute-value equations.**
          2. **Solve absolute-value inequalities.**

State/Local Objectives _____

_____

✓ Check the items you wish to use for this lesson.

STARTING OPTIONS
____ Homework Check: TE page 349; Answer Transparencies
____ Warm-Up or Daily Homework Quiz: TE pages 353 and 351, CRB page 51, or Transparencies

TEACHING OPTIONS
____ Motivating the Lesson: TE page 354
____ Lesson Opener (Graphing Calculator): CRB page 52 or Transparencies
____ Graphing Calculator Activity with Keystrokes: CRB page 53
____ Examples: Day 1: 1–2, SE page 353; Day 2: 3–5, SE pages 354–355
____ Extra Examples: Day 1: TE page 354 or Transp.; Day 2: TE pages 354–355 or Transp.; Internet
____ Technology Activity: SE page 359
____ Closure Question: TE page 355
____ Guided Practice: SE page 356; Day 1: Exs. 1, 2, 6–11; Day 2: Exs. 3–5, 12–18

APPLY/HOMEWORK
Homework Assignment
____ Basic   Day 1: 20–44 even; Day 2: 46–66 even, 69–71, 76–85
____ Average   Day 1: 20–44 even; Day 2: 46–68 even, 69–71, 76–85
____ Advanced   Day 1: 20–44 even; Day 2: 46–68 even, 69–73, 76–85

Reteaching the Lesson
____ Practice Masters: CRB pages 54–56 (Level A, Level B, Level C)
____ Reteaching with Practice: CRB pages 57–58 or Practice Workbook with Examples
____ Personal Student Tutor

Extending the Lesson
____ Applications (Real-Life): CRB page 60
____ Challenge: SE page 358; CRB page 61 or Internet

ASSESSMENT OPTIONS
____ Checkpoint Exercises: Day 1: TE page 354 or Transp.; Day 2: TE pages 354–355 or Transp.
____ Daily Homework Quiz (6.4): TE page 358, CRB page 64, or Transparencies
____ Standardized Test Practice: SE page 358; TE page 358; STP Workbook; Transparencies

Notes _____

_____

_____

TEACHER'S NAME _____ CLASS _____ ROOM _____ DATE _____

Lesson Plan for Block Scheduling

1-day lesson (See *Pacing the Chapter,* TE pages 330C–330D)   **For use with pages 353–359**

GOALS
1. **Solve absolute-value equations.**
2. **Solve absolute-value inequalities.**

State/Local Objectives _____

_____

_____

✓ **Check the items you wish to use for this lesson.**

STARTING OPTIONS
____ Homework Check: TE page 349; Answer Transparencies
____ Warm-Up or Daily Homework Quiz: TE pages 353 and
     351, CRB page 51, or Transparencies

TEACHING OPTIONS
____ Motivating the Lesson: TE page 354
____ Lesson Opener (Graphing Calculator): CRB page 52 or Transparencies
____ Graphing Calculator Activity with Keystrokes: CRB page 53
____ Examples 1–5: SE pages 353–355
____ Extra Examples: TE pages 354–355 or Transparencies; Internet
____ Technology Activity: SE page 359
____ Closure Question: TE page 355
____ Guided Practice Exercises: SE page 356

APPLY/HOMEWORK
Homework Assignment
____ Block Schedule: 20–68 even, 69–71, 76–85

Reteaching the Lesson
____ Practice Masters: CRB pages 54–56 (Level A, Level B, Level C)
____ Reteaching with Practice: CRB pages 57–58 or Practice Workbook with Examples
____ Personal Student Tutor

Extending the Lesson
____ Applications (Real-Life): CRB page 60
____ Challenge: SE page 358; CRB page 61 or Internet

ASSESSMENT OPTIONS
____ Checkpoint Exercises: TE pages 354–355 or Transparencies
____ Daily Homework Quiz (6.4): TE page 358, CRB page 64, or Transparencies
____ Standardized Test Practice: SE page 358; TE page 358; STP Workbook; Transparencies

Notes _____

_____

_____

CHAPTER PACING GUIDE	
Day	**Lesson**
1	Assess Ch. 5; 6.1 (all)
2	6.2 (all); 6.3 (all)
3	**6.4 (all)**
4	6.5 (all)
5	6.6 (all)
6	6.7 (all)
7	Review/Assess Ch. 6

Available as
a transparency

NAME _____ DATE _____

WARM-UP EXERCISES

For use before Lesson 6.4, pages 353–359

Which values of x make the statement true?

1. $|x| = 5$

2. $-|x| = -9$

Solve.

3. $x - 7 < -2$ or $x - 7 > 7$

4. $10 \geq 2x + 4 \geq -10$

DAILY HOMEWORK QUIZ

For use after Lesson 6.3, pages 346–352

1. Write an inequality that represents the statement "x is at most -4 or at least 1."

Solve the inequality. Write a sentence that describes the solution.

2. $-11 < -3x - 2 < 1$

3. $-2x - 3 > 5$ or $3x + 2 \geq -1$

Solve the inequality. Graph the solution. Is the indicated value of x a solution?

4. $3 \leq 2x - 7 < 11;\ x = 9$

5. $5x - 3 \leq 6$ or $7 < 4x - 9;\ x = 1.9$

NAME _____ DATE _____

Graphing Calculator Lesson Opener

For use with pages 353–358

In Questions 1–4, graph each equation using a standard viewing window.

1. Enter the equation $y = |x - 1|$ into your calculator.

   a. Estimate the x-intercept of the graph.

   b. Use the *Table* feature to check your estimate.

   c. Substitute the x-intercept into the equation $|x - 1| = 0$.
      What do you notice?

2. Enter the equation $y = |x| - 1$ into your calculator.

   a. Estimate the x-intercepts of the graph.

   b. Use the *Table* feature to check your estimate.

   c. Substitute the x-intercepts into the equation $|x| - 1 = 0$.
      What do you notice?

3. Enter the equation $y = |2x|$ into your calculator.

   a. Estimate the x-intercept of the graph.

   b. Use the *Table* feature to check your estimate.

   c. Substitute the x-intercept into the equation $|2x| = 0$.
      What do you notice?

4. Enter the equation $y = |2x| - 2$ into your calculator.

   a. Estimate the x-intercepts of the graph.

   b. Use the *Table* feature to check your estimate.

   c. Substitute the x-intercepts into the equation $|2x| - 2 = 0$.
      What do you notice?

Algebra 1
Chapter 6 Resource Book

NAME _____ DATE _____

Graphing Calculator Activity Keystrokes

For use with Technology Activity 6.4 on page 359

TI-82

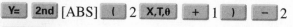

Y= 2nd [ABS] (2 X,T,θ + 1) − 2

ENTER

5 ENTER

ZOOM 6

2nd [CALC] 5 ENTER ENTER

Use the cursor keys, ◀ and ▶, to move the trace cursor to point of intersection at $x = -4$.

Press ENTER.

2nd [CALC] 5 ENTER ENTER

Use the cursor keys, ◀ and ▶, to move the trace cursor to select the other point of intersection at $x = 3$.

Press ENTER.

TI-83

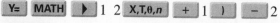

Y= MATH ▶ 1 2 X,T,θ,n + 1) − 2

ENTER

5 ENTER

ZOOM 6

2nd [CALC] 5 ENTER ENTER (-) 4

ENTER

2nd [CALC] 5 ENTER ENTER 3

ENTER

SHARP EL-9600c

Y= MATH [B] 1

2 X/θ/T/n + 1 ▶ − 2 ENTER

5 ENTER

ZOOM [A]5

2ndF [CALC] 2

2ndF [CALC] 2

CASIO CFX-9850GA PLUS

From the main menu, choose GRAPH.

OPTN F5 F1

(2 X,θ,T + 1) − 2 EXE

5 EXE SHIFT F3 F3 EXIT

F6 F5 F5

Press ▶ to find the next point of intersection.

Algebra 1
Chapter 6 Resource Book

NAME _____ DATE _____

Practice A
For use with pages 353–358

Use mental math to solve each equation.

1. $|x| = 8$

2. $|x| = 4$

3. $|x| = 6.5$

4. $|x| = -3$

5. $|x| + 2 = 3$

6. $|x| - 4 = 3$

Solve the equation.

7. $|x - 2| = 3$

8. $|x + 1| = 7$

9. $|x + 5| = 11$

10. $|x - 4| = 0$

11. $|2x - 1| = 5$

12. $|5x + 10| = 30$

13. $|3x - 6| = 21$

14. $|4x + 3| = 35$

15. $|3x + 8| = 7$

Complete the sentence using the word *and* or the word *or*.

16. $|x - 1| < 3$ means $x - 1 < 3$ _?_ $x - 1 > -3$

17. $|x - 1| > 3$ means $x - 1 > 3$ _?_ $x - 1 < -3$

18. $|x + 5| \le 6$ means $x + 5 \le 6$ _?_ $x + 5 \ge -6$

19. $|x + 5| \ge 6$ means $x + 5 \ge 6$ _?_ $x + 5 \le -6$

Solve the inequality.

20. $|x + 1| < 6$

21. $|x + 4| > 7$

22. $|x - 2| \ge 10$

23. $|2x - 3| \le 9$

24. $|3x + 6| > 15$

25. $|4x - 8| < 8$

Solve the inequality. Then graph the solution.

26. $|x + 5| \ge 3$

27. $|x + 3| < 17$

28. $|x - 7| \le 4$

29. $|3x - 9| > 6$

30. $|2x + 8| < 10$

31. $|5x - 1| \ge 2$

32. **Body Temperature**   Physicians consider an adult's body temperature to be normal if it is 98.6°F, plus or minus 1°F. Write an absolute value inequality that describe this normal temperature range.

33. **Music CD Prices**   The average price for a music CD is $15.50. Depending on where you shop, the price may vary by as much as $3.00. Write an absolute value inequality describing the possible prices of music CD's.

34. **Shoe Sizes**   The average size 8 shoe is approximately $8\frac{1}{2}$ inches long, plus or minus a quarter of an inch. Write an absolute value inequality that show the range of possible lengths of a person's foot that could wear a size 8.

35. **Car Mileage**   Your car averages 35 miles per gallon on the highway. The actual mileage varies from the average by 5 miles per gallon. Write an absolute value inequality that shows the range for the mileage your car gets.

NAME _____ DATE _____

Practice B

For use with pages 353–358

Solve the equation.

1. $|x - 4| = 2$

2. $|x + 3| = 9$

3. $|x - 5| = 8$

4. $|x + 4| = 1$

5. $|2x - 3| = 5$

6. $|4x - 5| = 11$

7. $|3x + 5| = 7$

8. $|6x + 3| = 21$

9. $|3x - 7| = 26$

10. $|x + 8.5| = 10.5$

11. $|x - 1.8| - 7 = 3$

12. $\left|x + \frac{1}{3}\right| = \frac{5}{3}$

Solve the inequality.

13. $|x + 2| < 5$

14. $|x + 4| > 9$

15. $|x - 3| \le 1$

16. $|3x - 6| > 3$

17. $|2x + 1| > 5$

18. $|2x - 3| \le 7$

19. $|x - 3.2| \le 8$

20. $|3x + 2| - 1 > 9$

21. $|x - 4| + 5 \le 7$

Solve the inequality. Then graph the solution.

22. $|x + 7| \ge 1$

23. $|x - 4| < 6$

24. $|x - 6| \ge 2$

25. $|4x - 5| < 11$

26. $|3x + 2| \le 8$

27. $|5x + 4| > 0$

28. *French Horn Range*   A French horn student has a range of no more than 17 notes from the middle C. Let $x = 0$ correspond to middle C. Write an absolute value inequality that shows the range of notes the student is able to play.

29. *Shampoo Prices*   The average price of a particular brand of shampoo is \$3.26. Depending on where you shop, the price may vary by as much as \$0.25. Write an absolute value inequality describing the possible prices of the shampoo. Solve the inequality.

30. *Tool and Die*   A tool and die shop makes a metal pull tab for a pop can. The length of the tab is 1.2 inches. This measurement may have an error of as much as 0.002 inches. Write an absolute value inequality that shows the range of possible lengths of the tabs. Solve the inequality.

31. *Elevation*   The highest elevation in North America is 20,320 feet *above* sea level at Mount McKinley. The lowest elevation is 282 feet *below* sea level in Death Valley. Find an absolute value equation that has the highest and lowest elevations in North America as its solution.

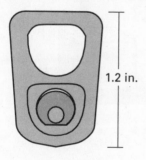

1.2 in.

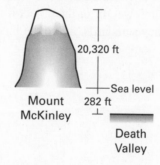

20,320 ft

Sea level

Mount   282 ft
McKinley

Death
Valley

NAME _____ DATE _____

Practice C

For use with pages 353–358

Solve the equation.

1. $|x - 7| = 4$ 2. $|x + 15| = 6$ 3. $|x + 3| = 18$

4. $|x - 20| = 36$ 5. $|3x + 6| = 39$ 6. $|7x - 4| = 3$

7. $|9 + 2x| = 7$ 8. $|4 - 6x| = 2$ 9. $\left|\frac{1}{4}x - 9\right| = 6$

10. $\left|\frac{3}{5}x + 2\right| = 11$ 11. $|20 - 3x| = 7$ 12. $|20 - 9x| = 5$

Solve the inequality. Then graph the solution.

13. $|x + 1| \leq 27$ 14. $|x - 8| > 14$ 15. $|6 - 5x| < 9$

16. $|x - 3| \leq 6$ 17. $|11 - x| < 20$ 18. $|7x + 3| < 11$

19. $\left|\frac{1}{4}x - \frac{1}{3}\right| \leq \frac{1}{3}$ 20. $|7 + 8x| > 5$ 21. $\left|18 + \frac{1}{2}x\right| \geq 10$

Write an absolute value inequality to fit the graph.

22.

23.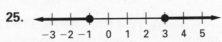

24.

25.

26. *Hourly Wages* The hourly wages at a local business are between \$8.60 and \$14.80 depending upon the job. Write an absolute value inequality describing the possible hourly wages of the employees.

27. *Attendance* The attendance at home basketball games has ranged between 820 and 1540 spectators. Write an absolute value inequality describing the possible attendance at home basketball games.

28. *Woodshop Class* In woodshop class, you must cut a piece of wood within $\frac{3}{16}$ inch of the teacher's specification of $5\frac{1}{8}$ inches. Write an absolute value inequality that describes the acceptable lengths for this piece of wood. Solve the inequality.

29. *Scale Accuracy* To test the accuracy of an industrial scale to 0.01 lb, an object that is known to be 100 lb is placed on the scale. Write an absolute value inequality that shows the measured weight of the object if the scale is *not* within accuracy standards. Solve the inequality.

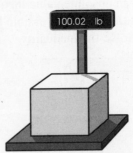

LESSON 6.4

Reteaching with Practice

For use with pages 353–358

GOAL Solve absolute-value equations and solve absolute-value inequalities

---

EXAMPLE 1 *Solving an Absolute-Value Equation*

Solve $|4x + 2| = 18$.

SOLUTION

Because $|4x + 2| = 18$, the expression $4x + 2$ can be equal to 18 or -18.

$4x + 2$ IS POSITIVE	$4x + 2$ IS NEGATIVE
$\|4x + 2\| = 18$	$\|4x + 2\| = 18$
$4x + 2 = +18$	$4x + 2 = -18$
$4x = 16$	$4x = -20$
$x = 4$	$x = -5$

The equation has two solutions: 4 and -5.

Exercises for Example 1

Solve the equation.

1. $|x| = 8$ **2.** $|x - 3| = 4$ **3.** $|2x - 3| = 9$

EXAMPLE 2 *Solving an Absolute-Value Inequality*

Solve $|x + 5| \le 1$.

SOLUTION

When an absolute value is *less than* a number, the inequalities are connected by *and*

$x + 5$ IS POSITIVE	$x + 5$ IS NEGATIVE
$\|x + 5\| \le 1$	$\|x + 5\| \le 1$
$x + 5 \le +1$	$x + 5 \ge -1$ ← Reverse inequality symbol.
$x \le -4$	$x \ge -6$

The solution is all real numbers greater than or equal to -6 *and* less than or equal to -4, which can be written as $-6 \le x \le -4$.

Reteaching with Practice

For use with pages 353–358

Exercises for Example 2

Solve the inequality.

4. $|x - 3| < 2$ **5.** $|8 + x| \le 3$ **6.** $|x - 1.5| < 1$

EXAMPLE 3   *Solving an Absolute-Value Inequality*

Solve $|2x - 1| > 5$.

SOLUTION

When an absolute value is *greater than* a number, the inequalities are connected by *or*.

$2x - 1$ IS POSITIVE	$2x - 1$ IS NEGATIVE
$\lvert 2x - 1 \rvert > 5$	$\lvert 2x - 1 \rvert > 5$
$2x - 1 > +5$	$2x - 1 < -5$ ← Reverse inequality symbol.
$2x > 6$	$2x < -4$
$x > 3$	$x < -2$

The solution of $|2x - 1| > 5$ is all real numbers greater than 3 *or* less than -2, which can be written as the compound inequality $x < -2$ *or* $x > 3$.

Exercises for Example 3

Solve the inequality.

7. $|x + 2| \ge 1$ **8.** $|x - 4| \ge 2$ **9.** $|2x + 1| > 3$

LESSON 6.4

Quick Catch-Up for Absent Students

For use with pages 353–359

The items checked below were covered in class on (date missed) _____

Lesson 6.4: Solving Absolute-Value Equations and Inequalities

____ **Goal 1:** Solve absolute-value equations. (p. 353)

Material Covered:

    ____ Example 1: Solving an Absolute-Value Equation

    ____ Example 2: Solving an Absolute-Value Equation

____ **Goal 2:** Solve absolute-value inequalities. (pp. 354–355)

Material Covered:

    ____ Activity: Investigating Absolute-Value Inequalities

    ____ Student Help: Study Tip

    ____ Example 3: Solving an Absolute-Value Inequality

    ____ Example 4: Solving an Absolute-Value Inequality

    ____ Example 5: Writing an Absolute-Value Inequality

Activity 6.4: Graphing Absolute-Value Equations (p. 359)

____ **Goal:** Solve an absolute-value equation using a graphing calculator.

    ____ Student Help: Keystroke Help

____ Other (specify) _____

_____

Homework and Additional Learning Support

    ____ Textbook (specify) <u>pp. 356–358</u>_____

_____

    ____ Internet: Extra Examples at www.mcdougallittell.com

    ____ *Reteaching with Practice* worksheet (specify exercises)_____

    ____ *Personal Student Tutor* for Lesson 6.4

Real Life Application:
When Will I Ever Use This?

Compact Disk (CD) Players

Many high school students listen to CDs and own CD players. CDs hold both music and other information for computers. The information and music is stored on CDs with a series of very small bumps in a long continuous spiral. The CD player needs to find and read back the data or bumps stored on the CD. The basic parts are a drive motor to spin the disk, a laser and lens, and a tracking mechanism. The player must be precisely calibrated or it will not be able to read the disk. The CD player must focus the laser on the track of bumps. The beam is bounced back to the lens. The bumps will reflect light one way while the smooth parts reflect another. The CD player interprets the reflected light as music or information.

One of the most difficult tasks is keeping the laser and lens focused. The player does this by using an autofocus signal based on the information that is being fed back to it.

In Exercises 1 and 2, use the following information.

A bump on a CD is generally between 1 and 3 microns long.

1. Write an absolute-value inequality describing the length of the bump.

2. Graph the inequality.

In Exercises 3 and 4, use the following information.

A CD can store between 73.9 and 74.1 minutes of music before it is considered full.

3. Write an absolute-value inequality describing the minutes of data a full CD will hold.

4. Graph the inequality from Exercise 3.

5. The plating of gold on a CD player part measures 0.56 microns thick. Does this fall within the required standards if a company uses $|x - 0.03|$ 0.525 for determining an acceptable part?

NAME _____ DATE _____

Challenge: Skills and Applications

For use with pages 353–358

In Exercises 1–2, solve the equation.

1. $|x - 3| = \dfrac{x}{4}$

2. $|5 - x| = x$

3. For what real numbers x is it true that $|x| \le x$? For what real numbers x is $|x| \ge x$?

In Exercises 4–6, solve the inequality.

4. $|x + 7| = x + 7$

5. $|n - 3| > n - 3$

6. $|4 - b| < 4 - b$

In Exercises 7–10, find the solution. Indicate if an equation is an *identity* or there is *no solution*.

7. $|5x| = 5|x|$

8. $|x + 3| = |x| + 3$

9. $|x| - 1 = |x| + 1$

10. $|x| + |-x| = 8$

In Exercises 11–14, use the following information.

Greg Jones is driving on a straight road that passes through the town of Westview. In Westview there is a radio station that can be heard anywhere within a 20-mile radius of the town. Greg started 50 miles away from Westview at noon, and he is driving at a rate of 40 miles/hour towards it.

11. Write and inequality, valid for the time before Greg reaches Westview, that states that his distance from Westview, expressed as a function of his driving time t (in hours), is less than or equal to 20 miles.

12. Write an inequality corresponding to the one from Exercise 11 for the time after Greg passes through Westview.

13. Combine the inequalities from Exercises 11 and 12 into one inequality using absolute value.

14. Solve the inequality from Exercises 13. Then find the time interval during which Greg will be able to hear the Westview radio station on his car radio.

TEACHER'S NAME _____ CLASS _____ ROOM _____ DATE _____

Lesson Plan

2-day lesson (See *Pacing the Chapter,* TE pages 330C–330D)   **For use with pages 360–367**

GOALS
1. **Graph a linear inequality in two variables.**
2. **Model a real-life situation using a linear inequality in two variables.**

State/Local Objectives _____

_____

✓ **Check the items you wish to use for this lesson.**

STARTING OPTIONS
____ Homework Check: TE page 356; Answer Transparencies
____ Warm-Up or Daily Homework Quiz: TE pages 360 and 358, CRB page 64, or Transparencies

TEACHING OPTIONS
____ Lesson Opener (Activity): CRB page 65 or Transparencies
____ Graphing Calculator Activity with Keystrokes: CRB page 66
____ Examples: Day 1: 1–4, SE pages 360–361; Day 2: 5, SE page 362
____ Extra Examples: Day 1: TE page 361 or Transp.; Day 2: TE page 362 or Transp.
____ Technology Activity: SE page 367
____ Closure Question: TE page 362
____ Guided Practice: SE page 363; Day 1: Exs. 1–12; Day 2: Exs. 13–14

APPLY/HOMEWORK
Homework Assignment
____ Basic   Day 1: 16–42 even; Day 2: 43–52, 57–64, 70, 71, 76, 78, 82, 88, 90; Quiz 2: 1–23
____ Average   Day 1: 16–42 even; Day 2: 43–52, 57–65, 70, 71, 76, 78, 82, 88, 90; Quiz 2: 1–23
____ Advanced   Day 1: 16–42 even; Day 2: 43–52, 57–65, 68, 70–73, 76, 78, 82, 88, 90; Quiz 2: 1–23

Reteaching the Lesson
____ Practice Masters: CRB pages 67–69 (Level A, Level B, Level C)
____ Reteaching with Practice: CRB pages 70–71 or Practice Workbook with Examples
____ Personal Student Tutor

Extending the Lesson
____ Applications (Interdisciplinary): CRB page 73
____ Challenge: SE page 365; CRB page 74 or Internet

ASSESSMENT OPTIONS
____ Checkpoint Exercises: Day 1: TE page 361 or Transp.; Day 2: TE page 362 or Transp.
____ Daily Homework Quiz (6.5): TE page 366, CRB page 78, or Transparencies
____ Standardized Test Practice: SE page 365; TE page 366; STP Workbook; Transparencies
____ Quiz (6.4–6.5): SE page 366; CRB page 75

Notes _____

_____

_____

TEACHER'S NAME _____ CLASS _____ ROOM _____ DATE _____

Lesson Plan for Block Scheduling

1-day lesson (See *Pacing the Chapter*, TE pages 330C–330D)   For use with pages 360–367

GOALS
1. **Graph a linear inequality in two variables.**
2. **Model a real-life situation using a linear inequality in two variables.**

State/Local Objectives _____

_____

_____

✓ **Check the items you wish to use for this lesson.**

STARTING OPTIONS
____ Homework Check: TE page 356; Answer Transparencies
____ Warm-Up or Daily Homework Quiz: TE pages 360 and
  358, CRB page 64, or Transparencies

TEACHING OPTIONS
____ Lesson Opener (Activity): CRB page 65 or Transparencies
____ Graphing Calculator Activity with Keystrokes: CRB page 66
____ Examples 1–5: SE pages 360–362
____ Extra Examples: TE pages 361–362 or Transparencies
____ Technology Activity: SE page 367
____ Closure Question: TE page 362
____ Guided Practice Exercises: SE page 363

APPLY/HOMEWORK
Homework Assignment
____ Block Schedule: 16–42 even, 43–52, 57–65, 70, 71, 76, 78, 82, 88, 90; Quiz 2: 1–23

Reteaching the Lesson
____ Practice Masters: CRB pages 67–69 (Level A, Level B, Level C)
____ Reteaching with Practice: CRB pages 70–71 or Practice Workbook with Examples
____ Personal Student Tutor

Extending the Lesson
____ Applications (Interdisciplinary): CRB page 73
____ Challenge: SE page 365; CRB page 74 or Internet

ASSESSMENT OPTIONS
____ Checkpoint Exercises: TE pages 361–362 or Transparencies
____ Daily Homework Quiz (6.5): TE page 366, CRB page 78, or Transparencies
____ Standardized Test Practice: SE page 365; TE page 366; STP Workbook; Transparencies
____ Quiz (6.4–6.5): SE page 366; CRB page 75

CHAPTER PACING GUIDE	
Day	Lesson
1	Assess Ch. 5; 6.1 (all)
2	6.2 (all); 6.3 (all)
3	6.4 (all)
4	**6.5 (all)**
5	6.6 (all)
6	6.7 (all)
7	Review/Assess Ch. 6

Notes _____

_____

_____

LESSON 6.5

NAME _____ DATE _____

WARM-UP EXERCISES

For use before Lesson 6.5, pages 360–367

Write the equation in slope-intercept form.

1. $x - 2y = -8$

2. $-y - x = 4$

Find the *x*- and *y*-intercepts of the graph of the equation.

3. $6x + 8y = 48$

4. $2.5x + 0.75y = 82.5$

....

DAILY HOMEWORK QUIZ

For use after Lesson 6.4, pages 353–359

Solve the equation or inequality.

1. $|x - 3| = 1$

2. $|3x - 7| = 8$

3. $|2x + 7| > 5$

4. Solve the inequality $|3 - 5x| \leq 7$. Then graph the solution.

5. Write an absolute-value inequality representing the possible values for a stock that had a low for the year of 23 and a high of 41.

NAME _____ DATE _____

Activity Lesson Opener

For use with pages 360–366

SET UP: Work in a group.
You will need graph paper.

1. Identify the inequality assigned to your group.

Inequality 1: $y \le -x + 2$ Inequality 2: $y > 2x - 1$

Inequality 3: $y < -x + 1$ Inequality 4: $y \ge x - 2$

2. As your teacher calls out x-coordinates, take turns
in your group and use your inequality to find a
corresponding y-coordinate. Plot the point on a
coordinate plane for your group.

$$-2 \qquad -1 \qquad 0 \qquad 1 \qquad -3$$

$$2 \qquad 3 \qquad \tfrac{1}{2} \qquad -\tfrac{1}{2} \qquad -4$$

3. By using their plotted points and their inequality,
each group can determine which of the graphs below
is the graph of their inequality.

A.
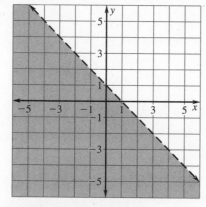

B.
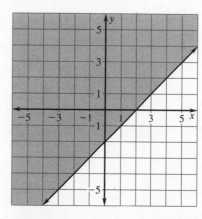

C.
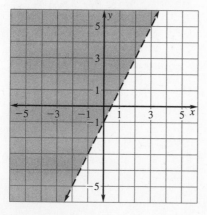

D.
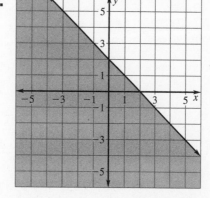

LESSON 6.5

Graphing Calculator Activity Keystrokes

For use with Technology Activity 6.5 on page 367.

TI-82

| WINDOW | ENTER |

| (-) | 10 | ENTER |

10 | ENTER |

1 | ENTER |

| (-) | 10 | ENTER |

10 | ENTER |

1 | ENTER |

| 2nd | [DRAW] 7

| X,T,θ | ÷ | 2 | + | 3 | , | 10 |) |

| ENTER |

TI-83

| Y= | X,T,θ,n | ÷ | 2 | + | 3 | ENTER |

Cursor to the left of Y1 and press | ENTER | to choose ◣ . (You may have to press | ENTER | several times.)

| ZOOM | 6

Sharp EL-9600c

| Y= | X/θ/T/n | ÷ | 2 | + | 3 | ENTER |

| 2ndF | [DRAW] [G] [1]

Press | − | to select Y1.

| ZOOM | [A]5

Casio CFX-9850Ga PLUS

From the main menu, choose GRAPH.

| F3 | F6 | F3 | X,θ,T | ÷ | 2 | + | 3 | EXE |

| SHIFT | F3 | F3 | EXIT | F6 |

NAME _____ DATE _____

Practice A

For use with pages 360–366

Check whether (0, 0) is a solution.

1. $x > -4$

2. $y > 3$

3. $x + y \le -3$

4. $x - y < 5$

5. $2x + y \ge 1$

6. $x - 2y < 2$

Is each ordered pair a solution of the inequality?

7. $x > 4$; $(3, 2), (-1, 4)$

8. $y \le 7$; $(7, 8), (9, 1)$

9. $y - x > 5$; $(3, -2), (4, 1)$

10. $x + y < -3$; $(-1, -5), (0, 0)$

11. $y - 2x \le 6$; $(-3, -1), (2, 7)$

12. $y + 4x \ge -2$; $(1, 3), (0, -2)$

13. $3x - y > 4$; $(6, 1), (-1, 1)$

14. $x - 2y \le -1$; $(3, 1), (4, 2)$

Sketch the graph of the inequality.

15. $y < 1$

16. $x \le -2$

17. $x + 5 \ge 8$

18. $y - 4 > 2$

19. $2x < 6$

20. $-3x \ge 12$

21. $5y \le -25$

22. $3 - x > 0$

23. $-4y < -16$

Sketch the graph of the inequality.

24. $y - x \ge 4$

25. $y + x \ge 2$

26. $y + x < -6$

27. $y - x < -2$

28. $y - 2x \ge 1$

29. $y + 3x \ge 2$

30. *Drama Club*   No more than 40 sophomores and juniors may participate in the drama club. Let *x* represent the number of sophomores and let *y* represent the numbers of juniors in the club. Write and graph an inequality that describes the different number of sophomores and juniors in the drama club.

31. *Perimeter*   The perimeter of a rectangle must be greater than 16 cm. Let *x* represent the length of the rectangle. Let *y* represent the width of the rectangle. Write and graph an inequality that describes the different lengths and widths of the rectangle.

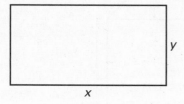

32. *Basketball*   In the last quarter of a high school basketball game, your team is behind by 26 points. A field goal is 2 points and a foul shot is 1 point. Let *x* represent the number of field goals scored. Let *y* represent the number of foul shots scored. Write and graph an inequality that models the different numbers of field goals and foul shots your team could score to win or tie the game. (Assume the other team scores no more points).

NAME _____ DATE _____

Practice B

For use with pages 360–366

Is each ordered pair a solution of the inequality?

1. $x + y < -1; (-3, -1), (0, 2)$

2. $y - x \geq 4; (-2, 2), (0, 4)$

3. $2x + 3y \leq 2; (5, 2), (2, -1)$

4. $4y - 2x > -3; (-1, 4), (6, 3)$

5. $5x - 3y \geq 15; (-4, -3), (2, 2)$

6. $6x + 2y < 24; (3, 2), (5, -3)$

7. $0.4x + 0.4y > 6.4; (-1, 18), (20, -2)$

8. $2.3x - 5.3y \leq 10; (-3, 0), (8, 5)$

9. $\frac{1}{2}x + \frac{1}{2}y < 7; (14, 0), (-21, -17)$

10. $\frac{1}{4}x - \frac{15}{24}y \geq \frac{1}{3}; (-10, -2), (1, 0)$

Sketch the graph of the inequality.

11. $x < 2$

12. $y \geq -4$

13. $x - 3 > -2$

14. $y + 5 \leq 7$

15. $-2x \geq -6$

16. $8y < 40$

17. $6 - x > 4$

18. $-6y < 30$

19. $3y < 1$

Sketch the graph of the inequality.

20. $x + y < 3$

21. $x - y \leq -5$

22. $-x - y > 4$

23. $y + 5x < -6$

24. $y - 3x > -2$

25. $3x - y \geq 2$

26. $2x + 3y > 12$

27. $\frac{1}{3}x + \frac{1}{2}y \leq 6$

28. $\frac{1}{4}x - \frac{3}{4}y < 7$

29. *Basketball* You're listening to the basketball game on your car radio. At half-time Collman has already scored 24 points, but you have to turn the car off and go to work. Let x represent the number of 2-point baskets scored. Let y represent the number of 3-point baskets scored. Write and graph the inequality that describes the different number of 2-point and 3-point shots Collman could have scored by the end of the game.

30. *Clothes* You have $125 to spend on school clothes. It costs $20 for a pair of pants and $15 for a shirt. Let x represent the number of pants you can buy. Let y represent the number of shirts you can buy. Write and graph an inequality that describes the different number of pants and shirts you can buy.

31. *Windows* The area of the window shown at the right is less than 28 square feet. Let x and y represent the height of the triangular and rectangular portions of the window. Write and graph an inequality that describes the different dimensions of the window.

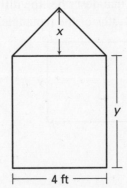

NAME _____ DATE _____

Practice C

For use with pages 360–366

Is each ordered pair a solution of the inequality?

1. $x + 2y \le -3$; $(0, 3)$, $(-5, 1)$

2. $5x - y > 2$; $(-5, 0)$, $(5, 23)$

3. $12x + 4y \ge 3$; $(1, -3)$, $(0, 2)$

4. $-8x - 3y < 5$; $(-1, 1)$, $(3, -9)$

5. $-x - y \le -10$; $(-3, -7)$, $(5, 4)$

6. $21x - 10y > 4$; $(2, 3)$, $(-1, 0)$

7. $\frac{1}{2}x - 7y \le 5$; $(20, 1)$, $(4, -2)$

8. $4x + 5y < 6$; $\left(\frac{1}{2}, 1\right)$, $\left(\frac{1}{2}, -1\right)$

Sketch the graph of the inequality.

9. $-x > 3$

10. $x \le -\frac{1}{3}$

11. $y \ge -4.8$

12. $y + 5.3 \le 7.8$

13. $x - 3.9 > -2.1$

14. $12y > -6$

15. $2x \ge -\frac{5}{2}$

16. $7x \le -\frac{7}{3}$

17. $8y < 16$

Sketch the graph of the inequality.

18. $y \ge 2x + 10$

19. $y < \frac{3}{4}x - 5$

20. $-6x + y \ge 3$

21. $7x - y < -3$

22. $2x + 3y \ge 6$

23. $4x + 12y > 3$

24. $x - 5y \le 15$

25. $7x - 7y > -21$

26. $4x - 2y \le 8$

27. $9x + 3y \le -1$

28. $2x + 4y \ge 16$

29. $\frac{2}{5}x + \frac{3}{4}y > 20$

30. *Math Club*   No more than 65 freshmen and sophomores may participate in the math club. There are three times as many freshmen as sophomores. Let x represent the number of freshmen. Let y represent the number of sophomores. Write and graph an inequality that describes the different number of freshmen and sophomores in the math club.

31. *Calculators*   A store carries $3000 worth of various types of calculators. A scientific calculator costs $20 and a graphing calculator costs $80. Let x represent the number of scientific calculators in stock. Let y represent the number of graphing calculators in stock. Write and graph the inequality that describes the different number of scientific and graphing calculators in stock.

32. *Breakfast*   You and your friends go to a bagel shop for breakfast. Together you have $12 to spend. Each bagel costs $.35 and each glass of juice costs $.80. Let x represent the number of bagels you can buy. Let y represent the number of juices you can buy. Write an inequality that represents the number of bagels x and the number of juices y that your group can afford. Sketch a graph of the inequality.

NAME _____ DATE _____

Reteaching with Practice

For use with pages 360–366

GOAL Graph a linear inequality in two variables and model a real-life situation using a linear inequality in two variables

---

VOCABULARY

A **linear inequality** in x and y is an inequality that can be written as $ax + by < c$, $ax + by \leq c$, $ax + by > c$, or $ax + by \geq c$.

An ordered pair (x, y) is a **solution** of a linear inequality if the inequality is true when the values of x and y are substituted into the inequality.

The **graph** of a linear inequality in two variables is the graph of the solutions of the inequality.

EXAMPLE 1 *Checking Solutions of a Linear Inequality*

Check whether the ordered pair is a solution of $3x - y \geq 2$.

a. $(0, 0)$ **b.** $(2, 0)$ **c.** $(2, 3)$

SOLUTION

(x, y)	$3x - y \geq 2$	Conclusion
a. $(0, 0)$	$3(0) - 0 = 0 \not\geq 2$	$(0, 0)$ is not a solution.
b. $(2, 0)$	$3(2) - 0 = 6 \geq 2$	$(2, 0)$ is a solution.
c. $(2, 3)$	$3(2) - 3 = 3 \geq 2$	$(2, 3)$ is a solution.

Exercises for Example 1
....

Is each ordered pair a solution of the inequality?

1. $x + 2y < 0$; $(0, 0)$, $(-1, -2)$ **2.** $2x + y > 3$; $(2, 2)$, $(-2, 2)$

EXAMPLE 2 *Graphing a Linear Inequality in Two Variables*

Sketch the graph of $x - y < 2$.

SOLUTION

The corresponding equation is $x - y = 2$. To graph this line, first write the equation in slope-intercept form: $y = x - 2$.

Graph the line that has a slope of 1 and a y-intercept of -2. Use a dashed line to show that the points on the line are not solutions.

The origin $(0, 0)$ is a solution and it lies above the line. So, the graph of $x - y < 2$ is all points above the line $y = x - 2$.

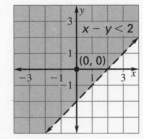

Reteaching with Practice

For use with pages 360–366

Exercises for Example 2

Sketch the graph of the inequality.

3. $x \le 2$　　　　**4.** $y > -1$　　　　**5.** $y - x < 3$　　　　**6.** $2x + y \ge 4$

EXAMPLE 3　*Modeling with a Linear Inequality*

You have \$16 to spend on crackers and cheese for an open house. Crackers cost \$2.50 per pound and cheese costs \$4 per pound. Let x represent the number of pounds of crackers you can buy. Let y represent the number of pounds of cheese you can buy. Write an inequality to model the amounts of crackers and cheese you can buy.

SOLUTION

Verbal Model	$\boxed{\text{Price of crackers}}$ · $\boxed{\text{Weight of crackers}}$ + $\boxed{\text{Price of cheese}}$ · $\boxed{\text{Weight of cheese}}$ $\le$ $\boxed{\text{Total cost}}$

Labels　　Price of crackers = 2.50　　(dollars per pound)

Weight of crackers = x　　　　　　(pounds)

Price of cheese = 4　　(dollars per pound)

Weight of cheese = y　　　　　　(pounds)

Total cost = 16　　　　　　(dollars)

Algebraic Model　　$2.50x + 4y \le 16$　　　Write algebraic model.

Exercise for Example 3

7. Graph the linear inequality in Example 3.

NAME _____ DATE _____

Quick Catch-Up for Absent Students

For use with pages 360–367

The items checked below were covered in class on (date missed) _____

Lesson 6.5: Graphing Linear Inequalities in Two Variables

_____ **Goal 1:** Graph a linear inequality in two variables. (pp. 360–361)

Material Covered:

_____ Example 1: Checking Solutions of a Linear Inequality

_____ Example 2: Graphing a Linear Inequality

_____ Example 3: Graphing a Linear Inequality

_____ Student Help: Study Tip

_____ Example 4: Writing in Slope-Intercept Form

Vocabulary:

linear inequality in x and y, p. 360   solution of a linear inequality, p. 360

graph of a linear inequality in   half-planes, p. 360
two variables, p. 360

_____ **Goal 2:** Model a real-life situation using a linear inequality in two variables. (p. 362)

Material Covered:

_____ Example 5: Modeling with a Linear Inequality

Activity 6.5: Graphing Inequalities (p. 367)

_____ **Goal:** Graph the solution of an inequality using a graphing calculator.

_____ Student Help: Keystroke Help

_____ Other (specify) _____

_____

Homework and Additional Learning Support

_____ Textbook (specify) <u>pp. 363–366</u>_____

_____

_____ *Reteaching with Practice* worksheet (specify exercises)_____

_____ *Personal Student Tutor* for Lesson 6.5

Interdisciplinary Application

For use with pages 360–366

Japan

SOCIAL STUDIES Japan is a country rich with a unique culture and identity. It is
known as a homogeneous, or uniform, country because it varies little by region,
religion, or ethnicity. This can be credited to several things. First, Japan is a
small island and was somewhat isolated for many centuries. Second, a large
population lives in this area and has lived under a centralized and relatively con-
trolling government for many centuries. Third, the Japanese culture places a lot
of emphasis on the group rather than the individual and this contributes to the
uniformity of its culture.

The Japanese are known for saving more of their disposable income than any
other industrialized country in the world. The Japanese money system is based
on the yen. One yen is the smallest unit of currency used, and is an aluminum
coin. The brass coin is equal to five yen and weighs 3.75 grams. The bronze coin
is ten yen and weighs 4.5 grams.

In Exercises 1 and 2, do the following.

Nina has been saving brass coins and bronze coins for several years in a large
container. She wants to know about how much money is in her container but she
doesn't want to count the coins. Instead she weighs the container and determines
that it weighs less than 1000 grams.

   1. Write an linear inequality describing the number of brass coins and bronze
       coins that may be in the container.

   2. Graph the inequality from Exercise 1.

In Exercises 3 and 4, use the following information.

Nina tells her friend, Ling, how she estimated the number of coins in her
container. Ling weighs his coins and determines that he has a little under
750 grams in his container.

   3. Write a linear inequality describing the number of coins Ling may have in
       his container.

   4. Graph the inequality from Exercise 3.

Challenge: Skills and Applications

For use with pages 360–366

In Exercises 1–2, write a linear inequality satisfying the given conditions.

Example: The points $(2, -3)$ and $(4, 1)$ are on the boundary of the graph of the inequality, and are solutions of the inequality. The point $(1, 5)$ is a solution of the inequality.

Solution: The slope of the boundary line is $\dfrac{1 - (-3)}{4 - 2} = 2$. Substituting $(4, 1)$ into the equation $y = 2x + b$ gives a value of -7 for b. Therefore, the equation of the boundary line is $y = 2x - 7$. Since $(1, 5)$ must satisfy the inequality, and since 5 is larger than $2(1) - 7$, the inequality sign is ">" or "≥." Because points $(2, -3)$ and $(4, 1)$ on the boundary line are solutions, the inequality is $y \geq 2x - 7$.

1. The points $(-3, 5)$ and $(7, 9)$ are on the boundary of the graph of the inequality and are solutions of the inequality. The origin is a solution of the inequality.

2. The boundary of the graph has slope $-\frac{1}{3}$ and passes through $(6, 7)$. The points on the boundary line are not solutions of the inequality. The point $(-3, 4)$ is a solution of the inequality.

In Exercises 3–6, use the following information.

Jean and Lisa Fisher have one phone in their home, which their parents let them use only between 8:00 P.M. and 9:30 P.M. Let x and y represent the numbers of minutes spent on the phone one evening by Jean and Lisa, respectively.

3. Write an inequality that models their parents' restriction on phone use.

4. Sketch the graph of the inequality.

5. How would the graph change if the sisters were given special permission to use the phone until 10:00 P.M.?

6. How would the graph change if Jean could still use the phone anytime from 8:00 P.M. until 9:30 P.M., but Lisa was required to be off the phone by 9:00 P.M.?

NAME _____ DATE _____

Quiz 2

For use after Lessons 6.4–6.5

1. Solve $|3x - 8| = 14$. *(Lesson 6.4)*

2. Solve $|7 - x| \geq 10$. *(Lesson 6.4)*

3. Is the ordered pair $(-4, 5)$ a solution of the inequality
$3x + 2y > 11$? *(Lesson 6.5)*

4. Sketch the graph of the inequality $-x - y < 9$. *(Lesson 6.5)*

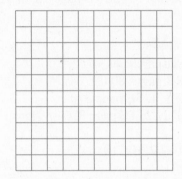

5. Write an inequality whose solution is shown in the graph.
(Lesson 6.5)

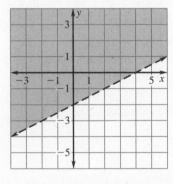

Answers

1. _____

2. _____

3. _____

4. Use grid at left.

5. _____

Lesson Plan

2-day lesson (See *Pacing the Chapter,* TE pages 330C–330D)   For use with pages 368–374

GOALS
1. **Make and use a stem-and-leaf plot to put data in order.**
2. **Find the mean, median, and mode of data.**

State/Local Objectives _____

_____

✓ **Check the items you wish to use for this lesson.**

STARTING OPTIONS
_____ Homework Check: TE page 363; Answer Transparencies
_____ Warm-Up or Daily Homework Quiz: TE pages 368 and 366, CRB page 78, or Transparencies

TEACHING OPTIONS
_____ Motivating the Lesson: TE page 369
_____ Lesson Opener (Application): CRB page 79 or Transparencies
_____ Graphing Calculator Activity with Keystrokes: CRB pages 80–81
_____ Examples: Day 1: 1–2, SE pages 368–369; Day 2: 3–4, SE page 370
_____ Extra Examples: Day 1: TE page 369 or Transp.; Day 2: TE page 370 or Transp.
_____ Closure Question: TE page 370
_____ Guided Practice: SE page 371; Day 1: Exs. 1–9; Day 2: Ex. 10

APPLY/HOMEWORK
Homework Assignment
_____ Basic   Day 1: 11–20; Day 2: 21–27, 29, 33, 36, 37, 41, 42, 44–46, 50–58 even
_____ Average   Day 1: 11–20; Day 2: 21–27, 29, 33, 34, 36, 37, 41, 42, 44–46, 50–58 even
_____ Advanced   Day 1: 11–20; Day 2: 21–27, 29, 30–34, 36, 37, 41–46, 50–58 even

Reteaching the Lesson
_____ Practice Masters: CRB pages 82–84 (Level A, Level B, Level C)
_____ Reteaching with Practice: CRB pages 85–86 or Practice Workbook with Examples
_____ Personal Student Tutor

Extending the Lesson
_____ Cooperative Learning Activity: CRB page 88
_____ Applications (Interdisciplinary): CRB page 89
_____ Challenge: SE page 374; CRB page 90 or Internet

ASSESSMENT OPTIONS
_____ Checkpoint Exercises: Day 1: TE page 369 or Transp.; Day 2: TE page 370 or Transp.
_____ Daily Homework Quiz (6.6): TE page 374, CRB page 93, or Transparencies
_____ Standardized Test Practice: SE page 374; TE page 374; STP Workbook; Transparencies

Notes _____

_____

_____

TEACHER'S NAME _____ CLASS _____ ROOM _____ DATE _____

Lesson Plan for Block Scheduling

1-day lesson (See *Pacing the Chapter*, TE pages 330C–330D)   For use with pages 368–374

GOALS   1. **Make and use a stem-and-leaf plot to put data in order.**
     2. **Find the mean, median, and mode of data.**

State/Local Objectives _____

_____

_____

✓ **Check the items you wish to use for this lesson.**

STARTING OPTIONS
____ Homework Check: TE page 363; Answer Transparencies
____ Warm-Up or Daily Homework Quiz: TE pages 368 and
          366, CRB page 78, or Transparencies

CHAPTER PACING GUIDE	
Day	**Lesson**
1	Assess Ch. 5; 6.1 (all)
2	6.2 (all); 6.3 (all)
3	6.4 (all)
4	6.5 (all)
5	**6.6 (all)**
6	6.7 (all)
7	Review/Assess Ch. 6

TEACHING OPTIONS
____ Motivating the Lesson: TE page 369
____ Lesson Opener (Application): CRB page 79 or Transparencies
____ Graphing Calculator Activity with Keystrokes: CRB pages 80–81
____ Examples 1–4: SE pages 368–370
____ Extra Examples: TE pages 369–370 or Transparencies
____ Closure Question: TE page 370
____ Guided Practice Exercises: SE page 371

APPLY/HOMEWORK
Homework Assignment
____ Block Schedule: 11–27, 29, 33, 34, 36, 37, 41, 42, 44–46, 50–58 even

Reteaching the Lesson
____ Practice Masters: CRB pages 82–84 (Level A, Level B, Level C)
____ Reteaching with Practice: CRB pages 85–86 or Practice Workbook with Examples
____ Personal Student Tutor

Extending the Lesson
____ Cooperative Learning Activity: CRB page 88
____ Applications (Interdisciplinary): CRB page 89
____ Challenge: SE page 374; CRB page 90 or Internet

ASSESSMENT OPTIONS
____ Checkpoint Exercises: TE pages 369–370 or Transparencies
____ Daily Homework Quiz (6.6): TE page 374, CRB page 93, or Transparencies
____ Standardized Test Practice: SE page 374; TE page 374; STP Workbook; Transparencies

Notes _____

_____

_____

NAME _____ DATE _____

WARM-UP EXERCISES

For use before Lesson 6.6, pages 368–374

Solve.

1. $x = \dfrac{98 + 92 + 88 + 95}{4}$

2. $x = \dfrac{3(35) + 3(27) + 2(28)}{3 + 3 + 2}$

3. $158 = \dfrac{x + 123 + 178}{3}$

DAILY HOMEWORK QUIZ

For use after Lesson 6.5, pages 360–367

1. Tell whether the ordered pairs $(1, -1)$ and $(4, 4)$ are solutions
of $3x - 2y > 5$.

Sketch the graph of the inequality.

2. $x + 3 \leq 5$

3. $x - y \geq 2$

4. $3y + 2x > 1$

Application Lesson Opener

For use with pages 368–374

The following data show the ages of science teachers at a local middle school.

54	24	38	21	48
43	59	62	31	40
37	52	36	25	48

1. Write the data in order from least to greatest.

2. Use the original list of data. Group the ages according to the tens digit. Then put each of those groups in order from least to greatest.

Tens Digit	*Data*
2	
3	
4	
5	
6	

3. Which was easier to order: the entire list of data or the groups of data? Why?

4. Use either ordered list of data. Which age occurs most often? Does this age tell you anything about the data?

5. How many ages are given? Which age is in the middle when the ages are listed in order? Does this age tell you anything about the data?

Lesson 6.6

NAME _____ DATE _____

Graphing Calculator Activity

For use with pages 368–374

GOAL **To use a graphing calculator to calculate the mean of a data set**

The *mean* of a data set is another name for the average. To calculate the mean, you find the sum of the numbers then divide by the number of items in the data set. If you enter the data into your graphing calculator, it will calculate the mean for you.

```
NAMES OPS MATH
1:min(
2:max(
3:mean(
4:median(
5:sum(
6:prod(
7↓stdDev(
```

Activity

1 Enter the following data into your graphing calculator in a list (L_1).

Ages of the students in your classroom: 13, 14, 14, 14, 14, 14, 14, 14, 15, 15, 15, 15, 15, 15, 15, 16, 16, 16, 16, 16

2 Use the *mean* function to calculate the mean age.

3 Does the mean appear to be a reasonable number to represent the set of data? Why or why not?

4 Add the age of your teacher, 47, to your list.

5 How do you think the addition of 47 to the data set will affect the mean?

6 Use the *mean* function to calculate the new mean age and check your guess.

7 Does the new mean appear to be a reasonable number to represent the set of data? Why or why not?

Exercises

1. Use the graphing calculator to find the mean of each data set.

   a. Bowling scores: 110, 125, 126, 126, 128, 130, 131, 131, 132, 135

   b. Number of siblings: 0, 1, 1, 1, 2, 2, 2, 2, 2, 3, 3, 3, 4, 4, 9

   c. Part-time annual incomes: 2000, 2050 2060, 2075, 3000, 3000, 3100, 3150

   d. Number of airplane flights: 0, 5, 5, 7, 8, 10, 10, 11, 12, 48

2. In which parts of Exercise 1 does the mean appear to be an *unreasonable* number to represent the data set? Explain your reasoning.

See page 81 for keystrokes.

NAME _____ DATE _____

Graphing Calculator Activity

For use with pages 368–374

TI-82

STAT 1

Enter the ages in L1.

13 ENTER 14 ENTER 14 ENTER
14 ENTER 14 ENTER 14 ENTER
14 ENTER 14 ENTER 15 ENTER
15 ENTER 15 ENTER 15 ENTER
15 ENTER 15 ENTER 15 ENTER
16 ENTER 16 ENTER 16 ENTER
16 ENTER 16 ENTER

2nd [QUIT] 2nd [LIST] ▶ 3 2nd
[L1])

ENTER

47 STO 2nd [L1] (21) ENTER
2nd [LIST] ▶ 3 2nd [L1])

TI-83

STAT 1

Enter the ages in L1.

13 ENTER 14 ENTER 14 ENTER
14 ENTER 14 ENTER 14 ENTER
14 ENTER 14 ENTER 15 ENTER
15 ENTER 15 ENTER 15 ENTER
15 ENTER 15 ENTER 15 ENTER
16 ENTER 16 ENTER 16 ENTER
16 ENTER 16 ENTER

2nd [QUIT] 2nd [LIST] ▶ ▶ 3
2nd [L1])

ENTER

47 STO 2nd [L1] (21) ENTER
2nd [LIST] ▶ ▶ 3 2nd [L1])

SHARP EL-9600c

STAT [A] ENTER

Enter the ages in L1.

13 ENTER 14 ENTER 14 ENTER
14 ENTER 14 ENTER 14 ENTER
14 ENTER 14 ENTER 15 ENTER
15 ENTER 15 ENTER 15 ENTER
15 ENTER 15 ENTER 15 ENTER
16 ENTER 16 ENTER 16 ENTER
16 ENTER 16 ENTER

2ndF [QUIT] 2ndF [LIST] [B] 3
2ndF [L1]) ENTER

47 STO 2ndF [L1] (21)

ENTER

2nd [LIST] 3 2ndF [L1])

ENTER

CASIO CFX-9850GA PLUS

From the main menu, choose STAT.
Enter the following in List 1.

13 EXE 14 EXE 14 EXE 14 EXE
14 EXE 14 EXE 14 EXE 14 EXE
15 EXE 15 EXE 15 EXE 15 EXE
15 EXE 15 EXE 15 EXE 16 EXE
16 EXE 16 EXE 16 EXE 16 EXE

F2 F1

The mean of List 1 is given on the first line to the
right of $\bar{x} =$.

EXIT

Use ▼ to cursor down to the 21st entry of List 1.

47 EXE

F1

The mean of List 1 is given on the first line to the
right of $\bar{x} =$.

Practice A

For use with pages 368–374

Make a stem-and-leaf plot for the data. Use the result to list the data in increasing order, in an ordered stem-and-leaf plot.

1. 28, 26, 32, 48, 36, 58, 44, 25, 42, 51, 50, 41, 37, 35

2. 18, 2, 26, 33, 22, 21, 5, 10, 11, 27, 3, 35, 29, 16, 14

3. 58, 61, 39, 55, 53, 32, 42, 67, 38, 43, 52, 41, 54, 61, 53

4. 17, 5, 2, 24, 31, 42, 30, 15, 56, 9, 31, 54, 34, 10, 16, 38, 59, 39, 23, 3

Find the mean, the median, and the mode of the collection of numbers.

5. 7, 6, 7, 8, 10, 8, 11, 8, 7, 7

6. 8, 13, 9, 8, 11, 8, 7, 6, 8, 12, 9

7. 42, 55, 41, 56, 67, 52, 41, 46, 58

8. 112, 115, 122, 121, 117, 119

9. 79, 85, 143, 113, 60, 146, 99, 171

10. 95, 111, 175, 46, 156, 297, 111, 218

Snowfall **In Exercises 11–13, use the following information.**

The table below shows the number of inches of snow that fell on 14 towns in a 50-mile radius during a snowstorm.

Town	A	B	C	D	E	F	G	H	I	J	K	L	M	N
Inches of Snow	8	4	7	6	5	6	7	8	9	10	11	5	4	8

11. Write the numbers in decreasing order.

12. Find the mean and the median for the set of data.

13. If another town in the area reported 20 inches of snow, would either the mean or the median change? Explain.

Winter Olympics **In Exercises 14–16, use the following information.**

The table shows the number of nations represented in the Winter Olympic Games from 1948 through 1994.

14. Make a stem-and-leaf plot of the data.

15. Find the mean, the median, and the mode for the set of data.

16. Which measure of central tendency do you think best represents the data? Explain.

Year	Nations
1948	28
1952	30
1956	32
1960	30
1964	36
1968	37
1972	35
1976	37
1980	37
1984	49
1988	57
1992	64
1994	67

NAME _____ DATE _____

Practice B

For use with pages 368–374

Make a stem-and-leaf plot for the data. Use the result to list the data in increasing order, in an ordered stem-and-leaf plot.

1. 21, 23, 41, 66, 52, 50, 37, 21, 20, 64, 66, 48, 53, 68, 39, 35

2. 2, 16, 27, 36, 2, 10, 19, 31, 32, 34, 25, 15, 23, 34, 20, 11, 8, 15

3. 24, 13, 17, 29, 11, 33, 20, 18, 37, 30, 39, 23, 10, 22, 28, 36, 22, 35, 15, 28

4. 18, 6, 52, 41, 43, 8, 29, 24, 33, 30, 2, 55, 28, 32, 8, 21, 5, 2, 38, 10, 54, 17

Find the mean, the median, and the mode of the collection of numbers.

5. 85, 90, 92, 91, 86, 90

6. 36, 37, 38, 38, 34, 35, 40, 39, 37, 36

7. 7, 8, 10, 12, 6, 8, 9, 8, 9, 11, 9, 8, 12

8. 52.8, 53.6, 53.9, 54, 54.5, 54.8, 55.1

9. 60, 62, 57, 59, 58, 60, 55, 60, 56

10. 42, 48, 43, 48, 51, 51, 48, 46, 46

Mountains   **In Exercises 11 and 12, use the following information.**

The table shows the world's 14 tallest mountains (in thousands of meters).

Aconcagua	7.0	Mt. Damávand	5.8
Annapurna	8.1	Mt. Everest	8.8
Cotopoxi	5.9	Mt. Godwin Austen (K-2)	8.6
Illampu	6.6	Mt. Logan	6.1
Kanchenjuga	8.6	Mt. Makalu	8.5
Kilimanjaro	5.9	Mt. McKinley	6.2
Lenin	7.1	Orizaba	5.7

11. Make a stem-and-leaf plot of the data.

12. Find the mean and median for the set of data.

Movie Prices   **In Exercises 13–15, use the following information.**

The table shows the average admission prices for movie theaters for selected years.

Year	1975	1980	1985	1990	1993	1994	1995	1996
Price	$2.05	$2.69	$3.55	$4.23	$4.14	$4.18	$4.35	$4.42

13. Make a stem-and-leaf plot of the data.

14. Find the mean, the median, and the mode for the set of data.

15. Which measure of central tendency do you think best represents the data? Explain.

NAME _____ DATE _____

Practice C
For use with pages 368–374

Make a stem-and-leaf plot for the data. Use the result to list the data in increasing order, in an ordered stem-and-leaf plot.

1. 35, 21, 85, 64, 28, 94, 45, 64, 33, 70, 47, 66, 62, 89, 42, 67, 47, 93, 91

2. 55, 84, 69, 59, 80, 73, 95, 55, 51, 63, 92, 57, 77, 60, 68, 56, 74, 83, 73, 64, 61

3. 123, 147, 140, 156, 133, 127, 139, 150, 141, 136, 144, 159, 137, 125, 136

4. 1.5, 2.7, 0.6, 3.1, 2.2, 4.7, 4.1, 3.5, 0.9, 1.6, 3.5, 2.2, 2.6, 4.7, 1.5, 3.7, 2.9

Find the mean, the median, and the mode of the collection of numbers.

5. 46, 25, 95, 36, 25, 74, 46, 43, 58

6. 340, 400, 225, 325, 255, 400, 400, 290

7. 1420, 1360, 1398, 1198, 1398, 1400

8. 952, 1022, 1062, 992, 925, 894, 925

9. 60.8, 56.7, 50.2, 60.8, 75.4, 97.2

10. 146.8, 158.4, 139.7, 147.5, 189.1, 116.1

Ice Cream **In Exercises 11 and 12, use the following information.**

The table shows the top 10 ice-cream consuming countries in the world (production per capita in pints).

United States	46.78
Finland	38.13
Denmark	34.76
Australia	32.64
Canada	29.22
Sweden	28.84
Norway	28.10
Belgium/Luxembourg	25.16
United Kingdom	21.96
New Zealand	21.87

11. Find the mean and the median for the set of data.

12. The mean is greater than the median. What does this suggest about the data set?

Speed **In Exercises 13–15, use the following information.**

The table shows the speed (in mph) of the winner of the Indianapolis 500 auto race in the years 1986–1997.

Year	Speed
1986	170.7
1987	162.2
1988	144.8
1989	167.6
1990	186.0
1991	176.5
1992	134.5
1993	157.2
1994	160.9
1995	153.6
1996	148.0
1997	145.9

13. Make a stem-and-leaf plot of the set of data.

14. Find the mean, the median, and the mode for the set of data.

15. Which measure of central tendency do you think best represents the data? Explain.

NAME _____ DATE _____

Reteaching with Practice

For use with pages 368–374

GOAL **Make and use a stem-and-leaf plot to put data in order and find the mean, median, and mode of data**

VOCABULARY

A **stem-and-leaf plot** is an arrangement of digits that is used to display and order numerical data.

A **measure of central tendency** is a number that is used to represent a typical number in a data set.

The **mean,** or **average,** of n numbers is the sum of the numbers divided by n.

The **median** of n numbers is the middle number when the numbers are written in order. If n is even, the median is the average of the two middle numbers.

The **mode** of n numbers is the number that occurs most frequently. A data set can have many modes or no mode.

EXAMPLE 1 *Making a Stem-and-Leaf Plot*

Make an ordered stem-and-leaf plot for the data.

16   8   35   2   22   10

31   50   13   35   56   28

SOLUTION

Use the digits in the tens' place for the stem and the digits in the ones' place for the leaves. Order the leaves to make an ordered stem-and-leaf plot. The key shows you how to interpret the digits.

Ordered stem-and-leaf plot

0	2	8		
1	0	3	6	
2	2	8		
Stem 3	1	5	5	Leaves
4				
5	0	6		Key: $3|1 = 31$

Reteaching with Practice

For use with pages 368–374

Exercises for Example 1

1. Use the stem-and-leaf plot from Example 1 to list the data in increasing order.

2. Make an ordered stem-and-leaf plot for the data and use the result to list the data in increasing order.

   16   7   38   19   11   26   2   33   27   39   2

EXAMPLE 2   *Finding the Mean, Median, and Mode*

Find the measure of central tendency of the data given in Example 1.

a. mean   **b.** median   **c.** mode

SOLUTION

a. To find the mean, add the 12 numbers and divide by 12.

$$\text{mean} = \frac{2 + 8 + 10 + 13 + 16 + 22 + 28 + 31 + 35 + 35 + 50 + 56}{12}$$

$$= \frac{306}{12}$$

The mean is 25.5.

b. To find the median, write the numbers in order and find the middle number. To order the numbers, use the ordered stem-and-leaf plot from Example 1.

2   8   10   13   16   22   28   31   35   35   50   56

Because $n = 12$ is even, the median is the average of the two middle numbers. The median is

$$\frac{22 + 28}{2} = 25.$$

c. To find the mode, use the ordered list in part (b). The mode is 35.

Exercise for Example 2

3. Find the mean, median, and mode of the data in Exercise 2.

NAME _____ DATE _____

Quick Catch-Up for Absent Students

For use with pages 368–374

The items checked below were covered in class on (date missed) _____

Lesson 6.6: Stem-and-Leaf Plots and Mean, Median, and Mode

____ **Goal 1:** Make and use a stem-and-leaf plot to put data in order. (p. 368)

Material Covered:

____ Example 1: Making a Stem-and-Leaf Plot

Vocabulary:

stem-and-leaf plot, p. 368

____ **Goal 2:** Find the mean, median, and mode of data. (pp. 369–370)

Material Covered:

____ Example 2: Finding the Mean, Median, and Mode

____ Student Help: Look Back

____ Example 3: A Bell-Shaped Distribution

____ Example 4: Interpreting Measures of Central Tendency

Vocabulary:

measure of central tendency, p. 369
mean, p. 369   average, p. 369
median, p. 369   mode, p. 369

____ Other (specify) _____

_____

Homework and Additional Learning Support

____ Textbook (specify) pp. 371–374 _____

_____

____ *Reteaching with Practice* worksheet (specify exercises) _____

____ *Personal Student Tutor* for Lesson 6.6

NAME _____   DATE _____

Cooperative Learning Activity

For use with pages 368–374

GOAL **To analyze data generated from a class survey**

Materials: paper, pencil

Exploring Survey Results

Measures of central tendency and certain types of graphs are used to
display data in a meaningful and easily understood way. In this activity,
you and a partner will survey the class on a topic that is of interest to
you, and then analyze and report the data collected from that survey

Instructions

1 Decide on a survey that is of interest
to you and your classmates. The
"answers" to your survey must be
in numerical form. For example, you
could survey your class on the number
of televisions each student has at home.

Stem	Leaf
0	1 2 2 2
1	0 1
2	3 4 7 9

2 Conduct your survey with the class.

3 Analyze the data by finding the mean,
median, and mode of your results.

4 Make a stem-and-leaf plot, a box-and-
whisker plot, and any other graph of
your choosing. It could be a bar, line,
or circle graph, or even a pictograph.

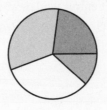

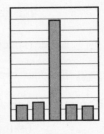

5 Report your results to the class, using
the statistics and graphs.

Analyzing the Results

1. Which measure of central tendency
best reports the average results of your
survey? Why?

2. Which graph best displays the results of your survey? What types
of information would best be reported using the other two graphs
you created?

NAME _____ DATE _____

Interdisciplinary Application

For use with pages 368–374

Pulse

PHYSICAL EDUCATION Pulse rate is the number of times per minute your heart beats. Each beat of the heart pushes blood into the lungs, where the blood replenishes its oxygen. Then the blood carries the oxygen to all of the muscles throughout the body. As a person's activity increases, his or her pulse will also increase. This occurs because the muscles of the body require more oxygen to support the increased level of activity. The larger the person, the slower the heart rate. A newborn baby's heart rate is about 120 beats per minute. The typical rate for an adult is about 72 beats per minute. Doctors consider resting rates from 60 to 100 beats per minute normal. Athletic training enlarges the heart and slows the heartbeat. Many well-trained athletes have resting rates from 40 to 60 beats per minute.

Your pulse can be determined by placing two fingers on your wrist (position fingers on the underside and thumb-side of your wrist). Count the number of beats in 30 seconds, and then double for the pulse rate per minute.

1. Have each student in your class determine his or her pulse. Collect the data.

2. Make a stem-and-leaf plot of the data in Exercise 1.

3. Use the stem-and-leaf plot in Exercise 2 to determine the mean, median, and mode of the data.

4. Have each student in your class jog in place for one minute and then take his or her pulse for 30 seconds and double the result. Collect the data.

5. Make a stem-and-leaf plot of the data in Exercise 4.

6. Use the stem-and-leaf plot in Exercise 5 to determine the mean, median, and mode of the data.

In Exercises 1–2, use the table that shows the sales of men's pants of various waist sizes at a certain store for one day.

Waist size	30	32	34	36	38	40	42	44
Number of pairs of pants	2	2	6	5	5	4	1	2

1. Find the mean, the median, and the mode of the waist sizes.

2. Suppose 3 more pairs of size 30 pants and 3 more pairs of size 40 pants were sold. Which of the three statistics you found in Exercise 1 will change and which will remain the same?

In Exercises 3–5, use the data below, which represent the annual salaries (in thousands of dollars) of the workers in a small company. Note that the amount of one salary, x, is missing from the data.

   25, 27, 27, 30, 32, 32, 50, 50, 50, 76, 100, x

3. Suppose you know the mode of the data with the missing salary included is 50, can you find the missing salary x? If so, what is it? If not, give any conclusions you can draw about x.

4. If the median of the data with x included is the same as it would be without x, what can you conclude about the value of x?

5. Suppose you know the mean of the data with x included is exactly $45,000. Can you find the exact value of x? If so, what is it? If not, explain why not.

In Exercises 6–9, use the stem-and-leaf plot.

In the mid-1990s, 38 states had annual revenues from tourism of less than $10 billion. The stem-and-leaf plot shows tourism revenue for a year for each of those 38 states.

0	8 8 9
1	3 3 4 5 6
2	0 2 3 5 7 8
3	0 0 1 6 9
4	0 1 4 4 5
5	0 8
6	1 4 6
7	0 1 5 7
8	0 8
9	2 4 8

Key 1|4 = $1.4 billion

6. Find the modes for the set of data.

7. Find the median for the set of data.

8. Find the mean for the set of data.

9. Find the median for all 50 states.

TEACHER'S NAME _____ CLASS _____ ROOM _____ DATE _____

Lesson Plan

2-day lesson (See *Pacing the Chapter,* TE pages 330C–330D)   For use with pages 375–382

GOALS   1. Draw a box-and-whisker plot to organize real-life data.
2. Read and interpret a box-and-whisker plot of real-life data.

State/Local Objectives _____

_____

✓ Check the items you wish to use for this lesson.

STARTING OPTIONS
_____ Homework Check: TE page 371; Answer Transparencies
_____ Warm-Up or Daily Homework Quiz: TE pages 375 and 374, CRB page 93, or Transparencies

TEACHING OPTIONS
_____ Motivating the Lesson: TE page 376
_____ Lesson Opener (Activity): CRB page 94 or Transparencies
_____ Graphing Calculator Activity with Keystrokes: CRB page 95
_____ Examples: Day 1: 1–2, SE pages 375–376; Day 2: 3, SE page 377
_____ Extra Examples: Day 1: TE page 376 or Transp.; Day 2: TE page 377 or Transp.; Internet
_____ Technology Activity: SE page 382
_____ Closure Question: TE page 377
_____ Guided Practice: SE page 378; Day 1: Exs. 1–8; Day 2: Exs. 9–10

APPLY/HOMEWORK
Homework Assignment
_____ Basic   Day 1: 11–22; Day 2: 25–36, 40–48 even, 49, 50; Quiz 3: 1–9
_____ Average   Day 1: 11–22; Day 2: 25–36, 40–48 even, 49, 50; Quiz 3: 1–9
_____ Advanced   Day 1: 11–22; Day 2: 25–37, 40–48 even, 49, 50; Quiz 3: 1–9

Reteaching the Lesson
_____ Practice Masters: CRB pages 96–98 (Level A, Level B, Level C)
_____ Reteaching with Practice: CRB pages 99–100 or Practice Workbook with Examples
_____ Personal Student Tutor

Extending the Lesson
_____ Applications (Real-Life): CRB page 102
_____ Challenge: SE page 380; CRB page 103 or Internet

ASSESSMENT OPTIONS
_____ Checkpoint Exercises: Day 1: TE page 376 or Transp.; Day 2: TE page 377 or Transp.
_____ Daily Homework Quiz (6.7): TE page 381 or Transparencies
_____ Standardized Test Practice: SE page 380; TE page 381; STP Workbook; Transparencies
_____ Quiz (6.6–6.7): SE page 381

Notes _____

_____

_____

Algebra 1
Chapter 6 Resource Book

TEACHER'S NAME _____ CLASS _____ ROOM _____ DATE _____

Lesson Plan for Block Scheduling

1-day lesson (See *Pacing the Chapter,* **TE pages 330C–330D)**   **For use with pages 375–382**

GOALS   1. Draw a box-and-whisker plot to organize real-life data.
2. Read and interpret a box-and-whisker plot of real-life data.

State/Local Objectives _____

_____

CHAPTER PACING GUIDE	
Day	**Lesson**
1	Assess Ch. 5; 6.1 (all)
2	6.2 (all); 6.3 (all)
3	6.4 (all)
4	6.5 (all)
5	6.6 (all)
6	**6.7 (all)**
7	Review/Assess Ch. 6

✓ **Check the items you wish to use for this lesson.**

STARTING OPTIONS
____ Homework Check: TE page 371; Answer Transparencies
____ Warm-Up or Daily Homework Quiz: TE pages 375 and
        374, CRB page 93, or Transparencies

TEACHING OPTIONS
____ Motivating the Lesson: TE page 376
____ Lesson Opener (Activity): CRB page 94 or Transparencies
____ Graphing Calculator Activity with Keystrokes: CRB page 95
____ Examples 1–3: SE pages 375–377
____ Extra Examples: TE pages 376–377 or Transparencies; Internet
____ Technology Activity: SE page 382
____ Closure Question: TE page 377
____ Guided Practice Exercises: SE page 378

APPLY/HOMEWORK
Homework Assignment
____ Block Schedule: 11–22, 25–36, 40–48 even, 49, 50; Quiz 3: 1–9

Reteaching the Lesson
____ Practice Masters: CRB pages 96–98 (Level A, Level B, Level C)
____ Reteaching with Practice: CRB pages 99–100 or Practice Workbook with Examples
____ Personal Student Tutor

Extending the Lesson
____ Applications (Real-Life): CRB page 102
____ Challenge: SE page 380; CRB page 103 or Internet

ASSESSMENT OPTIONS
____ Checkpoint Exercises: TE pages 376–377 or Transparencies
____ Daily Homework Quiz (6.7): TE page 381 or Transparencies
____ Standardized Test Practice: SE page 380; TE page 381; STP Workbook; Transparencies
____ Quiz (6.6–6.7): SE page 381

Notes _____

_____

_____

NAME ————————————————————— DATE ————

WARM-UP EXERCISES

For use before Lesson 6.7, pages 375–382

In its first seven games, your football team scored the following numbers of points in the first two quarters:
Quarter 1: 14, 10, 10, 17, 21, 16, 13
Quarter 2: 6, 7, 14, 13, 3, 15, 10

1. What was the median number of points scored in the first quarter?

2. What was the median number of points scored in the second quarter?

3. What was the median number of points scored in either quarter?

DAILY HOMEWORK QUIZ

For use after Lesson 6.6, pages 368–374

Make a stem-and-leaf plot of the data. Find the mean, median, and mode.

1. High temperatures (°F)

| 55 | 47 | 52 | 55 | 59 | 61 | 70 |
| 56 | 48 | 45 | 54 | 63 | 57 | 62 |

2. 100 meter race results (sec)

11.35	11.42	11.19	11.56	11.69
11.61	11.56	11.58	11.48	11.28
11.62	11.67	11.57	11.39	11.37

Lesson 6.7

SET UP: Work with a partner.
You will need blank index cards.

1. Write each number below on a card.

   14   8   22   16   19   9   25   12   5   11   10

2. Arrange the cards so the numbers are in increasing order. Place the cards in a row on a desk or table so the least number is on the left and the greatest number is on the right.

3. Start at the ends of the row of cards. Pick up one card in your left hand and one card in your right hand. Place those cards in a row directly under their position in the original row. Continue to pick up one card in each hand and move them down until you cannot pick up two cards. What does the number on the card remaining in the top row represent?

4. Start at the ends of the first section of the second row of cards. Use the procedure from Step 3 to begin a third row. What does the number on the card remaining in the first section of the second row represent? HINT: Think about how the number on the remaining card relates to the numbers in only the first section of the row.

5. Start at the ends of the second section of the second row of cards. Use the procedure from Step 3 to begin a third row. What does the number on the card remaining in the second section of the second row represent?

6. How many sections are in the third row? What do you notice about the number of cards in each section of the third row?

Graphing Calculator Activity Keystrokes

For use with Technology Activity 6.7 on page 382.

TI-82

STAT 1

Enter data in L1.

32 ENTER 24 ENTER 39 ENTER 55

ENTER 67 ENTER 35 ENTER 23 ENTER

61 ENTER 11 ENTER 44 ENTER 43

ENTER 30 ENTER 29 ENTER

2nd [STAT PLOT] 1

Choose the following.

On; Type: ⊞ ; Xlist: L1; Freq: 1

WINDOW ENTER 0 ENTER 70 ENTER 5

ENTER 0 ENTER 10 ENTER 1 ENTER

TRACE

Use the cursor keys, ◀ and ▶ , to find the quartile values.

TI-83

STAT 1

Enter data in L1.

32 ENTER 24 ENTER 39 ENTER 55

ENTER 67 ENTER 35 ENTER 23 ENTER

61 ENTER 11 ENTER 44 ENTER 43

ENTER 30 ENTER 29 ENTER

2nd [STAT PLOT] 1

Choose the following.

On; Type: ⊞ ; Xlist: L1; Freq: 1

ZOOM [A] 9 TRACE

Use the cursor keys, ◀ and ▶ , to find the quartile values.

SHARP EL-9600c

STAT [A] ENTER

Enter data in L1.

32 ENTER 24 ENTER 39 ENTER 55

ENTER 67 ENTER 35 ENTER 23 ENTER

61 ENTER 11 ENTER 44 ENTER 43

ENTER 30 ENTER 29 ENTER

2nd [STAT PLOT] [A] ENTER

Choose the following.

On; DATA X; ListX: L1

2nd [STAT PLOT] [E] 1

ZOOM [A] 9 TRACE

Use the cursor keys, ◀ and ▶ , to find the quartile values.

CASIO CFX-9850GA PLUS

From the main menu, choose STAT.

Enter data in L1.

32 EXE 24 EXE 39 EXE 55 EXE 67 EXE

35 EXE 23 EXE 61 EXE 11 EXE 44 EXE

43 EXE 30 EXE 29 EXE

F1 F6

Choose the following.

Graph Type: MedBox; Xlist: List1;

Frequency: 1; Outliers: Off ▫

EXIT

SHIFT F3 0 EXE 70 EXE 5 EXE 0 EXE

10 EXE 1 EXE EXIT

F1 F1 SHIFT F1

Use the cursor keys, ◀ and ▶ , to find the quartile values.

Practice A

For use with pages 375–381

Match the data with the box-and-whisker plot.

a.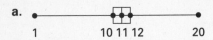

c.

b.

d.

1. 2, 1, 8, 13, 6, 8, 9, 15, 20, 18

2. 6, 7, 3, 1, 6, 11, 20, 3, 2, 1

3. 14, 15, 1, 8, 18, 20, 17, 19, 13, 15

4. 11, 10, 11, 12, 1, 13, 11, 10, 11, 20

Find the first, second, and third quartiles of the data.

5. 16, 5, 8, 9, 14, 11, 7, 13

6. 72, 78, 65, 94, 86, 80, 76, 90

7. 23, 29, 35, 21, 26, 30

8. 45, 58, 64, 57, 49, 53, 61

Draw a box-and-whisker plot of the data.

9. 3, 6, 25, 30, 29, 25, 19, 14, 20, 11

10. 75, 41, 44, 20, 15, 52, 64, 67, 48, 68, 48

11. Total drop of the highest waterfalls (in meters): 580, 561, 774, 657, 739, 979, 800, 610, 948, 646

Presidents **In Exercises 12–15, use the following information.**

The data shows the age of the Presidents of the United States at the time of their inauguration.

12. Make a stem-and-leaf plot to order the data.

13. Find the first, second, and third quartiles of the data.

14. Make a box-and-whisker plot of the data.

15. What does the plot tell you about the age of the Presidents?

57	61	57	57	58	57
61	54	68	51	49	64
50	48	65	52	56	46
54	49	50	47	55	55
54	42	51	56	55	54
51	54	51	60	62	43
55	56	61	52	69	64
46					

Exercise **Use the box-and-whisker plot that shows the amount of time (in hours) that adults spent exercising last week.**

16. What was the median amount of time exercising?

17. Did the same number of adults spend 3–6 hours exercising as adults exercising 6–12 hours? Explain.

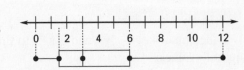

NAME _____ DATE _____

Practice B

For use with pages 375–381

Match the data with the box-and-whisker plot.

a.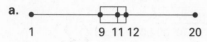
    1   9 11 12   20

b.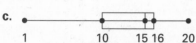
    1   5 5.5   7   20

c.
    1   10   15 16   20

d.
    1 2   5.5   8   20

1. 1, 10, 15, 18, 16, 14, 15, 16, 8, 20   **2.** 6, 2, 3, 8, 1, 2, 5, 6, 8, 20

3. 1, 12, 15, 8, 9, 10, 11, 12, 11, 20   **4.** 5, 8, 7, 6, 5, 4, 5, 6, 1, 20

Draw a box-and-whisker plot of the data.

5. 11, 21, 38, 0, 25, 31, 46, 42, 37, 15

6. 42, 37, 25, 49, 13, 26, 53, 42, 39, 24, 55, 27, 18, 31, 44, 11, 21, 33

7. Land area of 20 states (in thousands of square miles): 591, 159, 104, 84, 82, 97, 87, 70, 147, 77, 111, 122, 71, 70, 97, 77, 267, 85, 68, 98

Create a collection of 12 numbers that could be represented by the box-and-whisker plot.

8.
   10   12   16 17   20

9.
   75   87 92 94   99

Presidents In Exercises 10–14, use the following information.

The data shows the age of the Presidents of the United States at the time of their inauguration.

10. Make a stem-and-leaf plot to order the data.

11. Find the first, second, and third quartiles of the data.

12. Make a box-and-whisker plot of the data.

13. What does the plot tell you about the age of the Presidents?

14. A person must be 35 years old to be elected president. What would a new data point of 35 do to the box-and-whisker plot?

57	61	57	57	58	57
61	54	68	51	49	64
50	48	65	52	56	46
54	49	50	47	55	55
54	42	51	56	55	54
51	54	51	60	62	43
55	56	61	52	69	64
46					

Study Times Use the box-and-whisker plot that shows the amount of time (in hours) that students spent studying last week.

15. What was the median amount of time spent studying?

16. Did the same number of students spend 0–4 hours studying as students studying 7–9 hours? Explain.

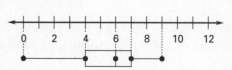

Match the data with the box-and-whisker plot.

a.

23.70   24.25 25.35   28.35   39.90

b.
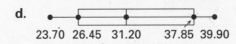
23.70 25.75   31.50 35.50   39.90

c.
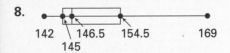
23.70   28.45 32.30 36.65 39.90

d.
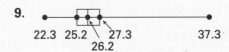
23.70 26.45 31.20   37.85 39.90

1. 33.7, 26.4, 36.8, 39.9, 29.3, 25.1, 34.2, 23.7

2. 27.1, 31.7, 23.7, 39.9, 35.2, 32.9, 29.8, 38.1

3. 23.7, 28.4, 25, 34, 38.4, 27.9, 37.3, 39.9

4. 24, 25.9, 23.7, 39.9, 29.1, 24.5, 24.8, 27.6

Draw a box-and-whisker plot of the data.

5. 79, 85, 36, 46, 55, 98, 44, 105, 67, 75

6. 6.5, 4.8, 2.0, 7.7, 9.8, 4.6, 12.7, 3.9, 4.5, 4.8, 3.7

7. Average annual snowfall in the ten snowiest cities (in inches): 100.8, 97.1, 116.1, 102.2, 240.8, 129.2, 110.0, 97.8, 114.0, 104.9

Create a collection of 12 numbers that could be represented by the box-and-whisker plot.

8.
142   145 146.5   154.5   169

9.
22.3 25.2   27.3   37.3
         26.2

Academy Awards **In Exercises 10–14, use the following information.**

The data shows the age of the actors and actresses who won the Academy Award for best actor and actress for the years 1987–1998.

10. Make a stem-and-leaf plot of the age of the actors.

11. Make a stem-and-leaf plot of the age of the actresses.

12. Make a box-and-whisker plot of the age of the actors.

13. Make a box-and-whisker plot of the age of the actresses.

14. Compare the two sets of data.

Year	Actor's Age	Actress's Age
1987	33	41
1988	51	25
1989	31	80
1990	42	42
1991	54	29
1992	51	33
1993	36	35
1994	37	45
1995	31	49
1996	45	39
1997	60	34
1998	45	25

Study Times **Use the box-and-whisker plot that shows the amount of time (in hours) that students spent studying last week.**

15. Which is greater, the mean or the median? Explain.

16. *True/False* More students studied less than 6 hours than studied more than 6 hours.

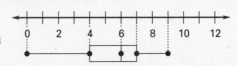

NAME _____ DATE _____

Reteaching with Practice

For use with pages 375–381

GOAL **Draw a box-and-whisker plot to organize real-life data and read and interpret a box-and-whisker plot of real-life data**

VOCABULARY

A **box-and-whisker plot** is a data display that divides a set of data into four parts.

The median or **second quartile** separates the set into two halves: the numbers that are below the median and the numbers that are above the median.

The **first quartile** is the median of the lower half.

The **third quartile** is the median of the upper half.

EXAMPLE 1 *Organizing Data*

The 1998 average monthly temperatures for your town are given below as ordered data.

33   35   40   42   47   52   54   57   64   66   67   72

a. Use the ordered data to find the quartiles.

b. Draw a box-and-whisker plot of the data.

SOLUTION

a. Second quartile: $\dfrac{52 + 54}{2} = 53$

First quartile: $\dfrac{40 + 42}{2} = 41$

Third quartile: $\dfrac{64 + 66}{2} = 65$

b. Draw a number line that includes the least number 33 and the greatest number 72 in the data set. Plot the least number, the quartiles, and the greatest number. Draw a line from the least number to the greatest number below your number line. Plot the same points on that line.

The "box" extends from the first to the third quartile. Draw a vertical line in the box at the second quartile. The "whiskers" connect the box to the least and greatest numbers.

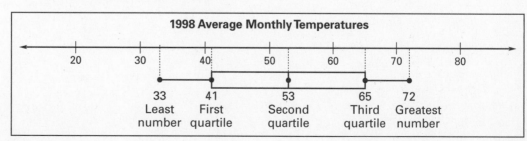

Algebra 1
Chapter 6 Resource Book

Reteaching with Practice

For use with pages 375–381

Exercise for Example 1

1. Draw a box-and-whisker plot of the ordered data.

    1   7   9   12   14   22   24   25

EXAMPLE 2 *Interpreting a Box-and-Whisker Plot*

The box-and-whisker plot below shows the number (in millions) of personal computers in the United States from 1985 to 1995.

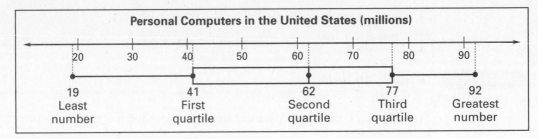

Personal Computers in the United States (millions)

| 19 | 41 | 62 | 77 | 92 |
| Least number | First quartile | Second quartile | Third quartile | Greatest number |

a. What is the median number of personal computers in the United States from 1985 to 1995?

b. What is the least number of personal computers in the United States from 1985 to 1995?

SOLUTION

a. The median number of personal computers in the United States from 1985 to 1995 is about 62 million.

b. The least number of personal computers in the United States from 1985 to 1995 is about 19 million.

Exercise for Example 2

2. What is the greatest number of personal computers in the United States from 1985 to 1995?

Algebra 1
Chapter 6 Resource Book

NAME _____ DATE _____

Quick Catch-Up for Absent Students

For use with pages 375–382

The items checked below were covered in class on (date missed) _____

Lesson 6.7: Box-and-Whisker Plots

____ **Goal 1:** Draw a box-and-whisker plot to organize real-life data. (pp. 375–376)

Material Covered:

____ Example 1: Finding Quartiles

____ Student Help: Study Tip

____ Example 2: Organizing Data

Vocabulary:

box-and-whisker plot, p. 375   second quartile, p. 375
first quartile, p. 375   third quartile, p. 375

____ **Goal 2:** Read and interpret a box-and-whisker plot of real-life data. (p. 377)

Material Covered:

____ Example 3: Interpreting a Box-and-Whisker Plot

Activity 6.7: Box-and-Whisker Plots (p. 382)

____ **Goal:** Draw a box-and-whisker plot using a graphing calculator or computer.

____ Other (specify) _____

_____

Homework and Additional Learning Support

____ Textbook (specify) <u>pp. 378–381</u>_____

_____

____ Internet: Extra Examples at www.mcdougallittell.com

____ *Reteaching with Practice* worksheet (specify exercises)_____

____ *Personal Student Tutor* for Lesson 6.7

NAME _____ DATE _____

Real Life Application:
When Will I Ever Use This?

For use with pages 375–381

Good Health and Test Scores

In your health class, the teacher is teaching a lesson on nutrition and states that eating a good breakfast and getting enough sleep will help you perform better in class. Some of your classmates are skeptical so the class decides to do an experiment. They break into three groups. The first group is going to sleep at least eight hours Thursday night and eat a nutritious breakfast the next morning. The second group eats no breakfast but gets at least eight hours of sleep. The third eats breakfast but gets four hours or less of sleep. Friday morning each group takes the same basic math and English test. The results are listed in the table.

Group 1	Group 2	Group 3
75, 82, 99, 81,	86, 83, 76, 91,	70, 85, 81, 69,
95, 96, 85, 79,	78, 81, 93, 96,	73, 87, 91, 80,
88, 94, 96, 89	85, 73, 90, 79	72, 69, 75, 76

1. Draw a stem-and-leaf plot to order each group of data.

2. Draw a box-and-whisker plot of each group.

3. Describe the results. Do they support the teacher's lesson?

Algebra 1
Chapter 6 Resource Book

NAME _____ DATE _____

Challenge: Skills and Applications

For use with pages 375–381

In Exercises 1–5, use the stem-and-leaf plot that shows the capacity of professional football stadiums, rounded to the nearest thousand.

6	0 0 0 1 1 2 3 4 5 5 6 6 6 7
7	0 0 1 1 3 3 3 4 5 6 8 8 9 9
8	0 0 Key 7|4 = 74,000

1. Draw a box-and-whisker plot of the capacities of the stadiums.

2. What conclusions can you draw about the data from the box-and-whisker plot?

3. Find the mode of the data. Which plot did you use? Why?

4. Find the median of the data. Which plot did you use? Why?

5. Find the mean of the data. Which plot did you use? Why?

In Exercises 6–7, use the table that shows the number of pupils per teacher by state, rounded to the nearest whole number.

Pupils per teacher	14	15	16	17	18	19	20	24
Frequency	6	7	10	13	4	4	4	2

6. Make a box-and-whisper plot of the data in the table.

7. What conclusions can you draw about the data from the box-and-whisker plot?

NAME _____ DATE _____

Chapter Review Games and Activities

For use after Chapter 6

Solve the following inequalities, and find the correct solution at the bottom of the page. Use the letter associated with the correct answer to answer the riddle:

What clothing item is worn by tornadoes?

1. $-15 < x - 7$

2. $x + 6 \geq -3$

3. $-25x \geq 75$

4. $9x + 4 \leq -6x - 26$

5. $|x + 5| > 7$

6. $|x - 8| + 3 \leq 11$

7. $8 - x < 1$

8. $-18 \leq 3x < 21$

9. $4x - 11 < 5$ or $8x - 7 > 9$

(F) -3

(S) -12 2

(O) 6

(D) -2

(I) -9

(G) -2

(A) -16 16

(K) -6 7

(T) -7 -6

(W) -8

(H) -7

(S) -8

(E) -2 2

(C) 7

(N) -3

(S)

(O) 0 16

(B) 2 4

$\overline{}$ $\overline{}$ $\overline{}$ $\overline{}$ $\overline{}$ $\overline{}$ $\overline{}$ $\overline{}$ $\overline{}$
1   2   3   4   5   6   7   8   9

Review and Assess

NAME _____ DATE _____

Chapter Test A

For use after Chapter 6

Graph the inequality.

1. $x \geq -3$

2. $x < 3.5$

Solve the inequality. Graph the solution on a number line.

3. $x - 3 \leq 1$

4. $a + 3 < 10$

5. You run a ten-kilometer race in 45.5 minutes. Write an inequality for the time of the runners who finished the race after you did.

Solve the inequality.

6. $3x + 2 \leq 17$

7. $2 - x > 5$

8. You are at the music store to buy some CDs. You have $45 to spend and the store sells CDs for $12.99 each. Write an inequality that represents the number of CDs that you can buy without spending more money than you have.

Write an inequality that represents the statement and graph the inequality.

9. x is less than 4 and greater than 1

Solve the inequality and graph the solution.

10. $6 < x + 4 \leq 11$

11. $x + 4 < 2$ or $x - 4 > -1$

Write a compound inequality that describes the graph.

12.

-5 -4 -3 -2 -1 0 1 2 3 4 5

13.

-3 -2 -1 0 1 2 3 4 5 6 7

1.	
2.	
3.	
4.	
5.	
6.	
7.	
8.	
9.	
10.	
11.	
12.	
13.	

Algebra 1
Chapter 6 Resource Book

105

Review and Assess

Solve the equation or the inequality.

14. $|x| = 5$

15. $|x + 2| < 5$

Is the ordered pair a solution of the inequality?

16. $x + y < 5; (3, 0)$

17. $x - y \geq 6; (2, 7)$

Sketch the graph of the inequality.

18. $x \leq 5$

19. $y > -3$

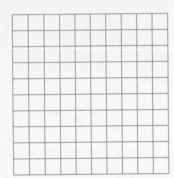

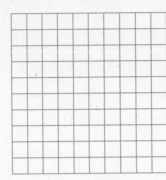

14.	_____
15.	_____
16.	_____
17.	_____
18.	_____
19.	_____
20.	_____
21.	_____
22.	_____

Make a stem-and-leaf plot of the data.

20. 40, 33, 20, 22, 36, 54, 27, 42, 30

Stem		Leaves

Find the mean, the median, and the mode of the collection of numbers.

21. 3, 1, 9, 5, 9, 6, 9

Find the first, second, and third quartiles of the data.

22. 6, 10, 1, 8, 3, 1, 4

Review and Assess

NAME _____ DATE _____

Chapter Test B

For use after Chapter 6

Solve the inequality. Graph the solution on a number line.

1. $p - 2 \geq -4$

2. $-y < 4$

3. You finish a two-mile walking race in 36.5 minutes. Write an inequality for the average speed of the walkers who finished after you did. (Average speed = distance/time)

Solve the inequality.

4. $4x - \frac{2}{3} \geq \frac{1}{3}$

5. $7 - 3x \leq 22$

6. The biology club budgeted $200 for their pancake breakfast. Each meal costs $1.50 to prepare. Write an inequality that represents the number of meals that can be prepared without going over the budget.

Write an inequality that represents the statement and graph the inequality.

7. x is greater than 1 or is less than -2

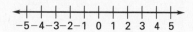

Solve the inequality and graph the solution.

8. $-3 \leq 2x + 5 < 11$

9. $4x + 5 < 3$ or $3x - 2 \geq 1$

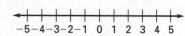

Write a compound inequality that describes the graph.

10.

-5-4-3-2-1 0 1 2 3 4 5

11.

-5-4-3-2-1 0 1 2 3 4 5

Solve the equation or the inequality.

12. $|x + 3| = 4$

13. $|x - 2| = 6$

14. $|x - 5| < 3$

15. $|2x + 3| \geq 17$

1.	_____
2.	_____
3.	_____
4.	_____
5.	_____
6.	_____
7.	_____
8.	_____
9.	_____
10.	_____
11.	_____
12.	_____
13.	_____
14.	_____
15.	_____

Review and Assess

Is the ordered pair a solution of the inequality?

16. $3x + 2y \le 4; (4, 3)$ **17.** $5x - 3y > 4; (-1, -5)$

Sketch the graph of the inequality.

18. $x + 4 > 5$ **19.** $y - 3 \le -4$

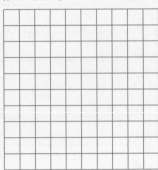   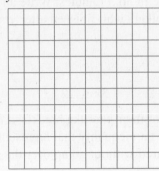

16. _____

17. _____

18. _____

19. _____

20. _____

21. _____

22. _____

23. _____

24. _____

20. You have $5 to spend on fruit for a picnic. Apples cost $0.99 per pound and bananas cost $0.49 per pound. Write an inequality to model the amounts of apples and bananas you can buy.

Make a stem-and-leaf plot of the data.

21. 54, 21, 34, 25, 51, 26, 45, 37, 31

Stem | Leaves

Find the mean, the median, and the mode of the collection of numbers.

22. 10, 7, 8, 7, 8, 8

Find the first, second, and third quartiles of the data.

23. 15, 6, 1, 13, 5, 11, 3, 8

Draw a box-and-whisker plot of the data.

24. 24, 16, 12, 28, 19, 21, 15

←┼┼┼┼┼┼┼┼┼┼┼┼┼┼┼┼┼→

Review and Assess

NAME _____ DATE _____

Chapter Test C

For use after Chapter 6

Solve the inequality. Graph the solution on a number line.

1. $-\dfrac{x}{5} \geq 3$

-18 -16 -14 -12 -10 -8

2. $-12b \leq 48$

```
<--+--+--+--+--+--+--+--+--+--+--+-->
  -5 -4 -3 -2 -1  0  1  2  3  4  5
```

3. You ran a ten-kilometer race in 50.5 minutes. Write an inequality for the average speed of the runners who finished after you did. (Average speed = distance/time)

Solve the inequality.

4. $2x + 5 < 3x - 7$

5. $-8x - 3 \geq -4x + 5$

6. An ice cream shop sells two scoops of ice cream for $1.50 and charges $0.70 for each additional scoop. You have $3.50 to spend. Write an inequality that represents the number of scoops of ice cream you can buy without spending more money than you have.

Write an inequality that represents the statement and graph the inequality.

7. x is greater than or equal to 5 or is less than 0

```
<--+--+--+--+--+--+--+--+--+--+--+-->
  -5 -4 -3 -2 -1  0  1  2  3  4  5
```

Solve the inequality and graph the solution.

8. $-4 < 2x + 5 \leq 12$

```
<--+--+--+--+--+--+--+--+--+--+--+-->
  -5 -4 -3 -2 -1  0  1  2  3  4  5
```

9. $10x - 4 \leq -24$ or $5x + 3 > 18$

```
<--+--+--+--+--+--+--+--+--+--+--+-->
  -5 -4 -3 -2 -1  0  1  2  3  4  5
```

Write a compound inequality that describes the graph.

10.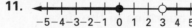
```
<--+--+--+--+--+--●--+--+--○--+-->
  -5 -4 -3 -2 -1  0  1  2  3  4  5
```

11.
```
<--+--+--+--+--+--●--+--+--○--+-->
  -5 -4 -3 -2 -1  0  1  2  3  4  5
```

Solve the equation or the inequality.

12. $|3x + 5| - 4 = 22$

13. $|2x - 7| - 2 \geq 11$

1. _____

2. _____

3. _____

4. _____

5. _____

6. _____

7. _____

8. _____

9. _____

10. _____

11. _____

12. _____

13. _____

Algebra 1
Chapter 6 Resource Book

Review and Assess

Is the ordered pair a solution of the inequality?

14. $\frac{2}{3}x + \frac{1}{3}y < 2; (-3, 5)$

15. $0.6x - 0.5y \geq 4; (-1, -1)$

Sketch the graph of the inequality.

16. $2x + y \leq 4$

17. $4x + 2y > 6$

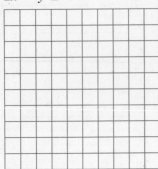

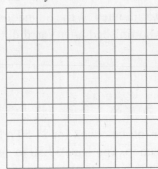

| 14. _____ |
| 15. _____ |
| 16. _____ |
| 17. _____ |
| 18. _____ |
| 19. _____ |
| 20. _____ |
| 21. _____ |
| 22. _____ |

18. An appliance store has $1000 to spend on stocking personal CD players and personal cassette players. The CD players cost $50 and the cassette players cost $20. Write an inequality to model the differ-ent numbers of CD players and cassette players the store can buy.

Make a stem-and-leaf plot of the data.

19. 23, 45, 55, 41, 23, 61, 57, 42, 22

Stem | Leaves

Find the mean, the median, and the mode of the collection of numbers.

20. 3.2, 1.5, 4.2, 2.5, 3.6, 4.8, 1.9

Find the first, second, and third quartiles of the data.

21. 6.4, 1.3, 3.9, 5.3, 4.2, 2.5, 3.6

Draw a box-and-whisker plot of the data.

22. 32, 20, 36, 19, 36, 27, 22, 23

NAME _____ DATE _____

SAT/ACT Chapter Test

For use after Chapter 6

1. Which inequality is equivalent to
   $-5x + 4 \leq -2x + 7$?

   Ⓐ $x \leq 1$ Ⓑ $x \geq 1$
   Ⓒ $x \leq -1$ Ⓓ $x \geq -1$

2. You are at a used book sale. Softcovers are
   $0.75 each and hardcovers are $1.50 each. If
   you have $6 to spend and you buy four soft-
   covers, how many hardcovers can you buy?

   Ⓐ 0   Ⓑ 1
   Ⓒ 2   Ⓓ 3

3. Which inequality represents the statement "x
   is less than 5 and is at least -5?"

   Ⓐ $-5 < x \leq 5$ Ⓑ $-5 \leq x \leq 5$
   Ⓒ $-5 < x < 5$ Ⓓ $-5 \leq x < 5$

4. Solve $-23 \leq 3x - 2 < 13$.

   Ⓐ $-7 \leq x < 5$ Ⓑ $-\frac{25}{3} \leq x < \frac{11}{3}$
   Ⓒ $-7 < x \leq 5$ Ⓓ $-\frac{25}{3} < x \leq \frac{11}{3}$

5. Which graph represents the solution of
   $6x - 4 \geq 14$ or $3x + 10 < 4$?

   Ⓐ
   ```
   ←+++++⊕+++++●++→
   -5-4-3-2-1 0 1 2 3 4 5
   ```

   Ⓑ
   ```
   ←+++++⊕+++++●++→
   -5-4-3-2-1 0 1 2 3 4 5
   ```

   Ⓒ
   ```
   ←+++++●+++++⊕++→
   -5-4-3-2-1 0 1 2 3 4 5
   ```

   Ⓓ
   ```
   ←+++++●+++++⊕++→
   -5-4-3-2-1 0 1 2 3 4 5
   ```

6. Solve $|8x + 2| - 4 = 18$.

   Ⓐ $\frac{3}{2}$ and -2 Ⓑ -2 and -3
   Ⓒ -3 and $\frac{5}{2}$ Ⓓ $-\frac{3}{2}$ and 2

7. Which ordered pair is *not* a solution of
   $5x + 4y < -12$?

   Ⓐ $(1, -5)$ Ⓑ $(-2, 4)$
   Ⓒ $(-4, 0)$ Ⓓ $(-3, -8)$

8. Choose the inequality whose solution is
   shown in the graph.

   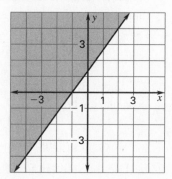

   Ⓐ $3y - 4x \geq 4$ Ⓑ $4x - 3y \geq 4$
   Ⓒ $3y - 4x \leq 4$ Ⓓ $4x - 3y \leq 4$

**In question 9, choose the statement
below that is true about the given numbers.**

   Ⓐ The number in column A is greater.
   Ⓑ The number in column B is greater.
   Ⓒ The two numbers are equal.
   Ⓓ The relationship cannot be determined
       from the given information.

9.

Column A	*Column B*
mean of 28, 16, 22, 13, 26	mean of 24, 10, 24, 30, 17

   Ⓐ   Ⓑ   Ⓒ   Ⓓ

Review and Assess

JOURNAL   **1.** You are talking to a friend who understands how to solve and graph a linear equation. You want to explain to your friend the differences in solving and graphing linear inequalities. (a) Explain how solving a linear *inequality* differs from solving a linear *equation*. (b) Explain how graphing a linear *inequality* in two variables is different from graphing a linear *equation* in two variables.

MULTI-STEP   **2.** Two television shows, A and B, took a random sampling of 25 viewers.
PROBLEM   Below is a box-and-whisker plot of the age of viewers of TV show A, and data for TV show B. Please note that an age of 0 means the viewer was less than a year old.

TV Show A:

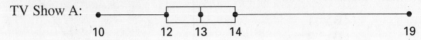

10   12   13   14   19

TV Show B:
(age of viewers): 11, 4, 7, 5, 15, 0, 3, 10, 4, 19, 6, 2, 5, 20, 6, 1, 12, 6, 14, 3,
                  0, 10, 15, 6, 4

a. Make an ordered stem-and-leaf plot for the age of viewers of TV show B.

b. Use the stem-and-leaf plot from part (a) to help you find the mean, the median, and the mode of the data. Which measure(s) of central tendency is most representative of the data? Explain why.

c. Draw a box-and-whisker plot of the TV Show B data. Using the same number line, copy the box-and-whisker plot of TV Show A below it.

d. Compare the number of viewers above 12 years of age for each TV show.

e. Write a compound inequality that represents the different ages of viewers of TV Show B.

f. Suppose a marketing department gears its advertisements towards children ages 14 to 19. Notice that the people who watch TV Show A range in age from 10 to 19. Does this mean that half of TV Show A's audience is within the targeted market? Why or why not?

g. Analyze the difference between the two sets of data. Be detailed and specific. For example, what type of market should TV Show A be geared toward? What about TV show B?

3. *Critical Thinking* Suppose the mean age of a viewer for TV Show A is 13.36 years. Is this a good representative measure of central tendency? Why or why not? Suppose one more viewer was interviewed and was found to be 45 years old. Find the new mean age of viewers of TV Show A. Do you think this new mean would be a representative measure of central tendency? Why or why not?

Alternative Assessment Rubric

For use after Chapter 6

**JOURNAL
SOLUTION**

1. Complete answer should address these points: **a.** In solving a linear inequality, the inequality symbol is reversed when dividing or multiplying both sides by a negative. **b.** For a linear inequality in two variables, the line is dashed for < or > and solid for ≤ or ≥. Also, one has to determine which way to shade. The solution of an inequality is a shaded region, whereas the solution of an equation is just a line.

**MULTI-STEP
PROBLEM
SOLUTION**

2. **a.**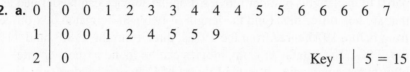

b. 7.52; 6; 6; median and mode; *sample explanation:* Most of the data seems to be clustered around 6.

c.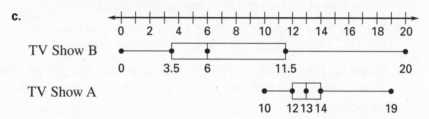

d. *Sample answer:* Less than 25% of TV Show B viewers are more than 12 years old, whereas 75% of TV Show A viewers are more than 12 years old.

e. $0 \leq x \leq 20$

f. no; 75% of the viewers are between 10 and 14 years old.

g. *Sample answer:* The ages of viewers of TV Show B is spread out, whereas the audience for TV Show A is more tightly grouped in the 12 to 14 year old range. TV Show A would be better for a product that is geared towards a specific age group, whereas TV Show B would be fine if advertisements are geared toward a wider age group. The median age of TV Show A viewers is 13 years, but only 6 years for TV Show B.

3. Yes, it's close to the median; 14.58 years; no; *sample explanation*: 45 years is not representative of the data and affects the mean by too much.

**MULTI-STEP
PROBLEM
RUBRIC**

4 Students complete all questions accurately. The stem-and-leaf plot is ordered, box-and-whiskers plots are drawn and spaced accurately. Students demonstrate an understanding that box-and-whisker plots are divided into quartiles.

3 Students complete the questions and explanations. The leaves of the stem-and-leaf plot may not be in order, or box-and-whisker plots may not be drawn to scale. Most questions are answered correctly, but explanations may not be thorough.

2 Students complete questions and explanations. Several errors may occur. Students do not understand that box-and-whiskers are divided into quartiles, and make inaccurate comparisons between the two data sets.

1 Work is incomplete. Solutions and explanations are incorrect. Plots are missing or inaccurate. Students do not seem to understand the data.

Review and Assess

Project: Not What They Used to Be

For use with Chapter 6

OBJECTIVE   Decide if popular movies made after 1970 are shorter than popular movies made before that date.

MATERIALS   paper, pencil, computer (optional)

INVESTIGATION   One movie critic claims that popular movies in the 1970s were very short compared to the lengths of older movies. Popular movies can be defined as ones that are box office hits. Find the length of two non-animated box office hits from before 1970, three from the 1970s, three from the 1980s, and three from the 1990s. The lengths of many movies can be found at the web site http://www.cinemedia.net or the Library of Congress catalog.

1. Write a compound inequality that represents the lengths of movies between the lengths of the two pre-1970 movies you found.

2. Find the mean length of the three 1970s movies you found. Do the same for the three 1980s movies and for the three 1990s movies.

3. Graph the inequality from Exercise 1 on a number line and plot the means from Exercise 2 on the same number line. Which means are solutions to the inequality?

4. Make a box-and-whisker plot for the 11 movie lengths you found. Are the movies from before 1970 in the upper quartile?

PRESENT YOUR RESULTS   Write a report presenting your results. Include your inequality, means, number line, and box-and-whisker plot. Discuss what the results imply for the lengths of popular movies in each decade. Were popular movies in the 1970s shorter than earlier movies? Were they in the 1980s? 1990s? Which graph gives a clearer answer, the number line or the box-and-whisker plot?

Project: Teacher's Notes

For use with Chapter 6

GOALS • Use a verbal model to write an algebraic equation or inequality to solve a real-life problem.

• Graph and compare real numbers using a number line.

• Write, solve, and graph compound linear inequalities and use compound linear inequalities to model and solve real-life problems.

• Find the mean, median, and mode of data and read and interpret box-and-whisker plots of real-life data.

• Make generalizations based on data.

MANAGING THE PROJECT Students may find the lengths of videotapes of popular movies easier than they can find the lengths of the original movies. Most of the videotape lengths can be found in the Library of Congress catalog or at the web site http://www.cinemedia.net.

One 1930s animated film, *Snow White and the Seven Dwarfs*, is in the top 50 box office hits of all time. However, this film is very short. You may want to discuss whether or not it is representative of films of the time. The project is more interesting if data are limited to non-animated films.

RUBRIC The following rubric can be used to assess student work.

4 The student writes and graphs an appropriate inequality, finds the means, and draws the box-and-whisker plot correctly. The report presents an insightful analysis of how the lengths of popular movies have changed over time and presents convincing evidence.

3 The student writes and graphs an appropriate inequality, finds the means, and draws the box-and-whisker plot. However, the student may not perform all calculations accurately or may not fully address the issues when interpreting the data. The report presents an analysis of the main question, but the presentation may not be as convincing as possible.

2 The student finds and analyzes data. However, the work may be incomplete or reflect misunderstanding. For example, the means may not be plotted on the number line or the box-and-whisker plot may not be drawn correctly or with an appropriate scale. The report may indicate a limited grasp of certain ideas or may lack key supporting evidence.

1 The inequality, graph, means, or box-and-whisker plot may be missing or may not show understanding of key ideas. The report does not indicate an opinion about changes in the lengths of popular movies over time or fails to support an opinion.

Review and Assess

Algebra 1
Chapter 6 Resource Book

115

Cumulative Review

For use after Chapters 1–6

Check whether the given number is a solution of the inequality. (1.4)

1. $2y - 21 < 0.9$; 8

2. $9 + 9x \geq 18$; 1

3. $14.2 \leq -(2x - 7)$; 20

4. $2k^5 - 100 \leq 89$; -7

5. $2z(56 - z) > 89$; -7

6. $\dfrac{8c - 9}{6} \leq 20$; 12

Write the numbers in increasing order. (2.1)

7. $8, -15, 10, \frac{13}{30}, -2.9, -\frac{100}{35}$

8. $-5.6, -5.05, -5.003, -0.506, 5.5, -5.069$

9. $3\frac{2}{5}, 3.25, -3, -3.023, -3.6, -3\frac{6}{7}$

10. $9, -\frac{4}{17}, -\frac{34}{9}, -8.9, -9.01, -1$

Find the sum of the matrices. (2.4)

11. $\begin{bmatrix} 0.69 & -9.6 \\ 0.5 & 0.02 \end{bmatrix} + \begin{bmatrix} 0.23 & -0.69 \\ -0.36 & 0.89 \end{bmatrix}$

12. $[-5\frac{1}{3} \quad 9\frac{8}{9} \quad -1] + [-5 \quad -9 \quad -1]$

13. $\begin{bmatrix} \frac{3}{2} \\ \frac{7}{6} \\ 1\frac{9}{2} \end{bmatrix} + \begin{bmatrix} -\frac{8}{15} \\ -2 \\ -3 \end{bmatrix}$

14. $\begin{bmatrix} 5 \\ -9.8 \end{bmatrix} - \begin{bmatrix} -6.5 \\ 9 \end{bmatrix} + \begin{bmatrix} -11 \\ 18 \end{bmatrix}$

Solve the equation. (3.1)

15. $-96 = 33 - y$

16. $\frac{30}{27} = s - \frac{1}{9}$

17. $r - 156\frac{2}{15} = -150\frac{1}{30}$

18. $y - (0.08) = 0.008$

19. $|-3| - 2x - 6 = 3$

20. $|-6.2| - (-3) = -5y$

Solve the equation if possible. (3.4)

21. $\frac{8}{5} - (-9n) = \frac{7}{2}n - 5$

22. $3.2y - 15.8 = 4.5y - 15.8$

23. $-\frac{1}{6}(6b) = \frac{1}{3}(6b - 30)$

24. $-5.6(2x - 26) = -(-14.6 - 15x)$

Find the x-intercept and the y-intercept of the line. Graph the equation. Label the points where the line crosses the axes. (4.3)

25. $y = x$

26. $6x - 2y = 24$

27. $y = 9x - 3$

28. $x - 6y = 13$

Write the equation in slope-intercept form. Find the slope and y-intercept. (4.6)

29. $6y = -15$

30. $2.6x - 9.6y = 18$

31. $4x + 3y - 12 = 0$

32. $x - y = 9$

33. $x - 3y = 21$

34. $2y - 0.5x = 0.5$

Cumulative Review

Write an equation of the line that is parallel to the given line and passes through the given point. (5.2)

35. $y = 9x + 6, (6, 9)$

36. $y = x - \frac{7}{8}, \left(-\frac{1}{2}, \frac{9}{2}\right)$

37. $y = -25x + 9, (-1, -1)$

38. $y = -\frac{11}{2}x - 12, (-9, -6)$

Write an equation in slope-intercept form of the line that passes through the given points. (5.3)

39. $(3, -2), (-3, 2)$

40. $(9, -5), (-6, 5)$

41. $(3, 0), (-3, 5)$

42. $\left(\frac{3}{8}, 1\right), \left(-\frac{1}{8}, -1\right)$

43. $\left(\frac{2}{5}, -\frac{5}{2}\right), \left(\frac{12}{5}, \frac{5}{6}\right)$

44. $(-5.6, 12), (6.4, 0)$

Write the standard form of the equation of the line passing through the given point with the given slope. (5.6)

45. $(-9, 0), m = -9$

46. $(-15, -6), m = 1.35$

47. $(0, 5), m = -\frac{1}{2}$

48. $(2, -3), m = -3.6$

49. $(-3, -3), m = -6$

50. $\left(-\frac{3}{8}, 9\right), m = 4$

Solve the inequality. (6.1, 6.2)

51. $t - 3t < 16$

52. $\frac{6}{5}x + 6 \le 48$

53. $-68.5y > 120.5$

54. $\frac{x}{16} \ge 48$

55. $s + 2 > 2(34 - s)$

56. $-8.3 \le -(1.8 - m)$

Solve the inequality. Graph the solution. (6.3)

57. $-7 \le 6x - 1 \le 53$

58. $6x + 5 < 8$ or $3x - 9 > 27$

59. $-5x - 7 > 3x + 9$

60. $-3 \le 6x - 1 < 3$

61. $-5x > 55$ or $8x > 64$

62. $-7x \ge 42$ or $4x \ge 12$

Solve the equation. (6.4)

63. $|x - 3| = 8$

64. $|7x| - 12 = 2$

65. $\left|s - \frac{6}{15}\right| = \frac{1}{30}$

66. $|y - 6.5| = 9.8$

Sketch the graph. (6.5)

67. $x \ge 6\frac{3}{4}$

68. $y > -5$

69. $x + y < 4$

70. $1 \le \frac{3}{2}x - y$

Make a stem-and-leaf plot for the data. Use the result to list the data in increasing order. (6.7)

71. 25, 36, 89, 12, 78, 22, 26

72. 12, 23, 14, 11, 19, 10

73. 73, 54, 87, 89, 72, 56

74. 100, 125, 168, 148, 152, 112

ANSWERS

Chapter Support

Parent Guide
Chapter 6

6.1: $x > -9$; no; yes

6.2: $5x - (3.84x + 8) \geq 224$; at least 200

6.3: $-4 \leq x < -3$ **6.4:** $|x - 250| \leq 5$

6.5: $0.22x + 0.34y \leq 10$; no; yes

6.6: mean 500.9, median 506.5, mode 512

6.7: 17, 19.5, 22

Prerequisite Skills Review

1. No **2.** No **3.** Yes **4.** Yes **5.** 10

6. $-\frac{3}{8}$ **7.** 7.417 **8.** $\frac{5}{3}$

9.
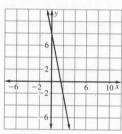

10.
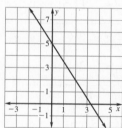

11.

12.
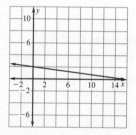

Strategies for Reading Mathematics

1. a. x is less than two times y **b.** x is greater than or equal to nine **c.** two times x is greater than negative four **d.** x minus eight is less than or equal to negative one **2.** The solid dot on the number line means that the number is included in the solution set. The open dot on the number line means that the number is not part of the solution set. **3. a.** Yes; negative two is greater than negative four. **b.** Yes; negative two plus four

equals two, which is equal to two. **c.** Yes; negative two minus four equals negative six, which is less than two. **d.** Yes; negative two plus three equals one, which is greater than zero.

4. Answers will vary. Example given. The point of the > or < symbols always points to the lesser number. The line under the inequality symbols means "or equal to." That line looks like one of the lines in an equal sign.

Lesson 6.1

Warm-up Exercises
1. -1 **2.** -3 **3.** 48 **4.** -4

Daily Homework Quiz

1. *Sample answer:* $y = 940x + 7000$

2. *Sample answer:* about 14,500 schools; linear extrapolation

Lesson Opener
Allow 10 minutes.

1. C; The amount Randall spent (a) is greater than the amount Karl spent ($25). **2.** A; The time Trevor studies (t) is less than the time Grace studies (1 hour). **3.** D; The number of coins Laura has (c) is at least as many (or greater than or equal to) the number of coins Lyle has (15).

4. B; The number of hours Nina and Drew work ($h + 4$) is at most (or less than or equal to) 10 hours.

Practice A

1. all real numbers greater than 5

2. all real numbers less than -4

3. all real numbers greater than or equal to 3

4. all real numbers less than or equal to -7

5. $x > 3$ **6.** $x < 1$ **7.** $x \leq -2$ **8.** $x \geq 5$

9.

10.

11.

12.

Lesson 6.1 *continued*

13.

14.

15. $x > 7$;

16. $x < 13$;

17. $x \leq 4$;

18. $x \leq -1$;

19. $x < -8$;

20. $x > -6$;

21. $x > 12$;

22. $x \leq -12$;

23. $x \leq 3$;

24. $x > -4$;

25. $x > -2$;

26. $x \leq -28$;

27. $s \geq 25$;

28. $A < 15$;

29. $T \leq -10$;

30. $P \leq 42$;

Practice B

1. $x < 1$ **2.** $x \geq 4$ **3.** $x \leq -2$ **4.** $x \geq -5$

5.

6.

7.

8.

9.
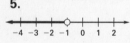

10.

11. $x < 4$;

12. $x \leq 7$;

13. $x > 2$;

14. $x \leq -4$;

15. $x < 2$;

16. $x < 4$;

17. $x \geq -6$;

18. $x < 6$;

19. $x \geq 3$;

20. $x > -9$;

21. $x \geq 18$;

22. $x > 4.5$;

23. $x \leq -5$;

24. $x \geq -6.1$;

25. $x > -11.5$;

26. $T > 98.6$;

27. $b > -268.9$;

28. $m \geq 32$;

29. $x \geq 26$; 26;
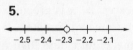

30. $E \geq -282$;
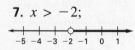

Practice C

1.

2.
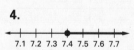

3.

4.
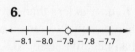

5.

6.
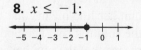

7. $x > -2$;

8. $x \leq -1$;

9. $x \leq 4$;

10. $x \leq -9$;

Lesson 6.1 *continued*

11. $x < 1$;

12. $x < -33$;

13. $x \geq 28$;

14. $x \geq 45$;

15. $x \geq \frac{1}{4}$;

16. $x > -5$;

17. $x \geq 16.5$;

18. $x \geq 3.2$;

19. $x > -7.5$;

20. $x \leq -14.4$;

21. $x > -5.2$;

22. $C \leq 46.95$;

23. $V < 95$;

24. $h \geq 75$;

25. $m < 3550$;

26. $x \geq 37$; \$37;

27. $s > 9\frac{29}{55}$;

Reteaching with Practice

1.

2.

3.

4. $2 > y$

5. $x \leq -3$

6. $k > -2$

7. $x < -4$

8. $a \leq 3$

9. $t < -6$

Real-Life Application

1. $h \leq 20$ **2.**

3. $s \leq 90$ **4.**

5. $h \geq 12$ **6.**

Challenge: Skills and Applications

1. $ac < bc$ **2.** $x \geq -5$ **3.** $x \leq 8$ **4.** $x \geq 3$
5. $x \leq 2$ **6.** Make sure students choose a and b so that $a > 0$ and $b = -a$.   **7.** Make sure students choose a and b so that $b = a - 1$.
8. Make sure students choose a and b so that $a < 0$ and $b = -3a$.   **9.** Make sure students choose a and b so that $b = a + 3$.   **10.** never
11. always   **12.** sometimes; *sample values:* true for $b = 2$, false for $b = \frac{1}{2}$.   **13.** always
14. sometimes; *sample values:* true for $a = \frac{1}{3}$ and $b = \frac{1}{2}$, false for $a = 2$ and $b = 3$.
15. never

Lesson 6.2

Warm-up Exercises

1. 4   **2.** 4   **3.** 0.2   **4.** -6

Daily Homework Quiz

1.

2. $x \leq 3$

3. $x > -3$

4. $x \geq -4.5$ **5.** $x < 42$

Lesson Opener

Allow 10 minutes.

1. The value of $2x + 3$ is equal to 5 when $x = 1$.
2. Any of the points chosen will result in a true statement.   **3.** Any of the points chosen will result in a false statement.   **4.** $x \geq 1$
5. The value of $5x - 6$ is equal to 9 when $x = 3$.
6. Any of the points chosen will result in a false statement.   **7.** Any of the points chosen will result in a true statement.   **8.** $x \leq 3$

Lesson 6.2 *continued*

Practice A

1. 6   2. $3\frac{1}{2}$ 3. -4 4. 6   5. 12   6. -3

7. $x < 6$ 8. $x < 5$ 9. $x \geq 5$ 10. $x \geq 4$

11. $x < -3$ 12. $x \geq 1$ 13. $x > 2$

14. $x > 3$ 15. $x \geq -1$ 16. $x > 2$

17. $x \leq -10$ 18. $x < 2$ 19. $y = 30x + 620$

20. $800 = 30x + 620$ 21. $30x + 620 < 800$;
$x < 6$ 22. $32x + 6 \leq 166$; $x \leq 5$; 5

23. 29 minutes   24. from 1981–1990

Practice B

1. $x < 7$ 2. $x > -5$ 3. $x > \frac{2}{3}$ 4. $x \leq 2$

5. $x > -7$ 6. $x \leq 2$ 7. $x \leq -\frac{1}{2}$ 8. $x > 1$

9. $x \leq 2$ 10. $x < -9$ 11. $x \leq 1$

12. $x < -\frac{1}{2}$ 13. $x \leq -\frac{1}{2}$ 14. $x > 6$

15. $x \geq 10$ 16. $x \leq \frac{4}{5}$ 17. $x < -2$

18. $x \leq 0$ 19. $y = 24x + 840$

20. $24x + 840 < 900$; $x < 2.5$

21. $12x + 5 \leq 50$ 22. $1.50x + 5 \leq 25$;
$x \leq 3.75$; 3 $x \leq 13\frac{1}{3}$; 13

23. from 1981–1990   24. $5x > 45$; $x > 9$

Practice C

1. $x > 7$ 2. $x > -5$ 3. $x \leq \frac{6}{7}$ 4. $x \leq -3$

5. $x < 2$ 6. $x > -15$ 7. $x \geq \frac{1}{3}$ 8. $x \geq 9$

9. $x > 3$ 10. $x > 3$ 11. $x < 0$ 12. $x \leq -\frac{3}{2}$

13. $x \leq 2$ 14. $x < 15$ 15. $x > -48$

16. $x \leq \frac{5}{2}$ 17. $x > \frac{1}{2}$ 18. $x \geq -\frac{1}{3}$ 19. 1997

20. 17 months   21. 64 minutes

22. $5x > 42$; $x > 8.4$

23. $\frac{1}{2} \cdot x \cdot 8 > 60$; $x > 15$

Reteaching with Practice

1. $x > 1$ 2. $m \leq 2$ 3. $y \geq -1$ 4. $a > -3$

5. $x \geq 6$ 6. $y \leq 6$ 7. at least 44 hours

8. at least 47 hours

Interdisciplinary Application

1. $3.5x + 40 \geq 0$ years since 1908

2. $29.6x + 108 \geq 0$ 3. $16.9x + 40 \geq 0$

Challenge: Skills and Applications

1. $x < 3$ 2. $a \leq 0$ 3. $w < -36$ 4. no
solution   5. $y > 16.5$ 6. all values of r

7. $7(n - 3) < 9(n - 5)$ 8. more than 12 tapes

9–10. Check students' inequalities. Samples are given.

9. *Sample inequality:* $3x - 4 \geq -1\frac{1}{2} + 2x$

10. *Sample inequality:* $4x - 3 < 5(x + 2) - 8$

Lesson 6.3

Warm-up Exercises

1. $-2, -1$ 2. $-5, -4, 2, 3, 4, 5$

3. $-5, -4, -3, -2$ 4. $0, 1, 2, 3, 4, 5$

Daily Homework Quiz

1. $x \leq -1$ 2. $x < 9$ 3. $x < 5$

4. $\frac{1}{2}x(15) \leq 105$; $x \leq 14$

Lesson Opener

Allow 10 minutes.

1. A; 2, 3, and 4 are all between 5 and 1.

2. C; -1, 0, 1, 2, and 3 are all integers between
-2 and 3, including 3.   3. D; the word "or"
means that the integers can be either less than
$-3(x < -3)$ or greater than $4(x > 4)$.

4. C; the only integer that is not less than 2 or
greater than 2 is 2.

Graphing Calculator Activity

1. a. $x < -1$; $x > 1$ b. $x < 0$; $x > 6$

c. $x > -2$; $x < 4$ 2. c.; *Sample answer:* $-1, 1$

Practice A

1. $0 < x < 4$

2. $x > 4$ or $x < 2$

3. $5 \leq x \leq 10$

4. $x < -3$ or $x > 5$

5. $x > -2$ or $x < -6$

6. $-3 \leq x < 4$

Lesson 6.3 *continued*

7. $6 \le x \le 10$ **8.** $x < 7$ or $x > 9$

9. $-3 < x \le 1$ **10.** $x \le -2$ or $x \ge 2$

11.
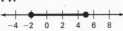

12.

13.

14.
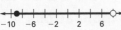

15.

16.

17. $-9 \le x < 8$

18. $-2 < x < 3$

19. $x < -8$ or $x \ge 6$

20. $-9 < x < -1$

21. $x \le -8$ or $x > 7$

22. $x \ge -5$ or $x < -16$

23. $-2 < x \le 9$

24. $x < -2$ or $x > 0$

25. $x < -3$ or $x > 6$

26. $15 < x < 20$ **27.** $98°F \le T \le 134°F$

Practice B

1. $-2 < x \le 1$ **2.** $x \le 8$ or $x \ge 9$

3. $x < -4$ or $x > 0$ **4.** $x \le 10$ or $x > 30$

5.

6.

7.

8.

9.

10.

11. $6 < x \le 8$;

12. $-6 < x < 0$;

---

13. $x > 4$ or $x < 0$;
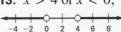

14. $-2 \le x \le 2$;

15. $-12 < x < -6$;
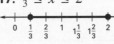

16. $x > 4$ or $x \le 1$;

17. $\frac{1}{3} \le x \le 2$
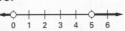

18. $-2 \le x < 4$

19. $x \ge 3$ or $x < -\frac{1}{2}$
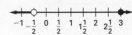

20. $-5 < x \le 5$; solution
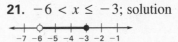

21. $-6 < x \le -3$; solution

22. $x \ge 4$ or $x \le 3$; solution

23. $x < -3$ or $x > \frac{4}{3}$; not a solution

24. $-70°F \le T \le 10°F$

25. a. 2 miles **b.** 6 miles **c.** $2 \le d \le 6$

Practice C

1. $-8 \le x \le -4$ **2.** $x < -\frac{1}{2}$ or $x > 1$

3. $x < 2.8$ or $x \ge 2.9$ **4.** $-5.1 \le x \le -4.8$

5.

6.

7.

8.

9.

10.

11. $-6 \le n \le -1$

12. $-\frac{3}{2} < x \le 3$
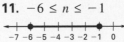

Answers

Lesson 6.3 *continued*

13. $x < -4 \text{ or } x > 0$

14. $-4 \le x \le 2$

15. $3 < x < 5$

16. $x \le 3 \text{ or } x \ge 6$

17. $-3 \le n < \frac{3}{2}$

18. $-1 < x < 1$

19. $x > -0.52 \text{ or } x < -5$

20. $-6 < x < 21$

21. $0.5 \le x < 2.5$

22. $x < 4 \text{ or } x > 6$

23. $45 \le x \le 65$

24. $50 \le x \le 70;\ 40 \le x \le 60$

25. $6.1 < x < 24.5$ **26.** $3 \le d \le 7, 5 \le d \le 9$

Reteaching with Practice

1. $-4 < x \le -2;$

2. $x > 3 \text{ or } x < -1;$

3. $-2 < x \le 3;$

4. $-3 < x < -1;$

5. $2 \ge x > -3;$

6. $x < 4 \text{ or } x \ge 5;$

7. $x > -1 \text{ or } x \le -3$

8. $172{,}000 \le v \le 226{,}000$

Real-Life Application

1. $40{,}368 < x < 74{,}560$

2. $18{,}737 \le x < 32{,}629$

3. $18{,}737 < x < 24{,}398$

4. $6000 \le x \le 30{,}000;\ 5000 \le x \le 15{,}000$

5.  ;

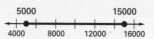

Math and History

1. $t \le 9$ **2.** $t < 3$ **3.** $t \ge 12$ **4.** $29 < t < 33$

5.

First pony express delivery in 1860.   First telephone was patented in 1876.

First fax in 1851.   Radio was invented in 1895.   First television was demonstrated in 1926.

1850 1860 1870 1880 1890 1900 1910 1920 1930

Challenge: Skills and Applications

1. $a \ge b$ **2.** $a > b$ **3.** $ac > cx > bc$

4. $\frac{3}{8} < x \le \frac{25}{8}$

5.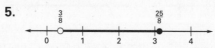

6. no   **7.** yes   **8.** $x \le -0.3 \text{ or } x > 0.9$

9.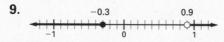

10. yes   **11.** no   **12.** 14 gal/min

13. 90 gal/min   **14.** $14 \le f \le 90$

Quiz 1

1. $x \ge 8;$

2. $x > -28;$

3. $h < 1000;$

4. $x > 11$ **5.** $48 \ge 4x$

6. $-4 \le x < 2$

7. $x \le 1 \text{ or } x > 4;$

Lesson 6.4

Lesson 6.4

Warm-up Exercises

1. 5 and -5 **2.** 9 and -9
3. $x < 5$ or $x > 14$ **4.** $-7 \leq x \leq 3$

Daily Homework Quiz

1. $x \leq -4$ or $x \geq 1$ **2.** $-1 < x < 3$; x is greater than -1 and less than 3.
3. $x < -4$ or $x \geq -1$; x is less than -4 or x is at least -1.
4. $5 \leq x < 9$;   no

5. $x \leq 1.8$ or $x > 4$;   no

Lesson Opener

Allow 10 minutes.
1. a. 1   **b.** From the table, when $y = 0$, $x = 1$.
c. It is a solution to the equation.   **2. a.** -1, 1
b. From the table, when $y = 0$, $x = 1$, and $x = -1$.   **c.** They are solutions to the equation.
3. a. 0   **b.** From the table, when $y = 0$, $x = 0$.
c. It is a solution to the equation.   **4. a.** -1, 1
b. From the table, when $y = 0$, $x = 1$, and $x = -1$.   **c.** They are solutions to the equation.

Practice A

1. $8, -8$ **2.** $4, -4$ **3.** $6.5, -6.5$
4. no solution   **5.** $1, -1$ **6.** $7, -7$ **7.** $5, -1$
8. $6, -8$ **9.** $6, -16$ **10.** 4   **11.** $3, -2$
12. $4, -8$ **13.** $9, -5$ **14.** $8, -\frac{19}{2}$
15. $-\frac{1}{3}, -5$ **16.** and   **17.** or   **18.** and   **19.** or
20. $-7 < x < 5$ **21.** $x < -11$ or $x > 3$
22. $x \leq -8$ or $x \geq 12$ **23.** $-3 \leq x \leq 6$
24. $x < -7$ or $x > 3$ **25.** $0 < x < 4$
26. $x \leq -8$ or $x \geq -2$ **27.** $-20 < x < 14$

28. $3 \leq x \leq 11$

29. $x < 1$ or $x > 5$

30. $-9 < x < 1$

31. $x \leq -\frac{1}{5}$ or $x \geq \frac{3}{5}$

32. $|t - 98.6| \leq 1.0$ **33.** $|p - 15.50| \leq 3.00$
34. $\left| x - 8\frac{1}{2} \right| \leq \frac{1}{4}$ **35.** $|x - 35| \leq 5$

Practice B

1. $6, 2$ **2.** $6, -12$ **3.** $13, -3$ **4.** $-3, -5$
5. $4, -1$ **6.** $4, -\frac{3}{2}$ **7.** $\frac{2}{3}, -4$ **8.** $3, -4$
9. $11, -\frac{19}{3}$ **10.** $2, -19$ **11.** $11.8, -8.2$
12. $\frac{4}{3}, -2$ **13.** $-7 < x < 3$
14. $x > 5$ or $x < -13$ **15.** $2 \leq x \leq 4$
16. $x < 1$ or $x > 3$ **17.** $x < -3$ or $x > 2$
18. $-2 \leq x \leq 5$ **19.** $-4.8 \leq x \leq 11.2$
20. $x < -4$ or $x > \frac{8}{3}$ **21.** $2 \leq x \leq 6$
22. $x \leq -8$ or $x \geq -6$ **23.** $-2 < x < 10$

24. $x \leq 4$ or $x \geq 8$ **25.** $-\frac{3}{2} < x < 4$

26. $-\frac{10}{3} \leq x \leq 2$ **27.** $x > -\frac{4}{5}$ or $x < -\frac{4}{5}$

28. $|x| \leq 17$ **29.** $|x - 3.26| \leq 0.25$
30. $|x - 1.2| \leq 0.002$, $1.198 \leq x \leq 1.2002$
31. $|x - 10,301| = 10,019$

Practice C

1. $3, 11$ **2.** $-9, -21$ **3.** $-21, 15$
4. $-16, 56$ **5.** $-15, 11$ **6.** $\frac{1}{7}, 1$ **7.** $-1, -8$
8. $\frac{1}{3}, 1$ **9.** $60, 12$ **10.** $15, -\frac{65}{3}$ **11.** $\frac{13}{3}, 9$
12. $\frac{5}{3}, \frac{25}{9}$
13. $-28 \leq x \leq 26$ **14.** $x < -6$ or $x > 22$

Lesson 6.4 *continued*

15. $-\frac{3}{5} < x < 3$

16. $-3 \le x \le 9$

17. $-9 < x < 31$

18. $-2 < x < \frac{8}{7}$

19. $0 \le x \le \frac{8}{3}$

20. $x < -\frac{3}{2}$ or $x > -\frac{1}{4}$

21. $x \le -56$ or $x \ge -16$

22. $|x - 8| \le 2$ **23.** $|x - 5| > 1$

24. $|x - 1| < 3$ **25.** $|x - 1| \ge 2$

26. $|x - 11.70| \le 3.10$ **27.** $|x - 1180| \le 360$

28. $\left|x - 5\frac{1}{8}\right| \le \frac{3}{16}, \ 4\frac{15}{16} \le x \le 5\frac{5}{16}$

29. $|x - 100| > 0.01, \ x < 99.99$ or $x > 100.01$

Reteaching with Practice

1. $8, -8$ **2.** $-1, 7$ **3.** $-3, 6$ **4.** $1 < x < 5$

5. $-11 \le x \le -5$ **6.** $0.5 < x < 2.5$

7. $x \le -3$ or $x \ge -1$ **8.** $x \le 2$ or $x \ge 6$

9. $x < -2$ or $x > 1$

Real-Life Application

1. $|\ell - 2| \le 1$

2.

3. $|m - 74| \le 0.1$

4.   **5.** no

Challenge: Skills and Applications

1. $4, \frac{12}{5}$ **2.** $\frac{5}{2}$ **3.** $x \ge 0$; all real numbers x

4. $x \ge -7$ **5.** $n < 3$ **6.** no solution

7. identity   **8.** $x \ge 0$ **9.** no solution

10. $4, -4$ **11.** $50 - 40t \le 20$

12. $40t - 50 \le 20$ **13.** $|50 - 40t| \le 20$

14. $\frac{3}{4} \le t \le \frac{7}{4}$; between 12:45 P.M. and 1:45 P.M.

Lesson 6.5

Warm-up Exercises

1. $y = \frac{1}{2}x + 4$ **2.** $y = -x - 4$

3. $8, 6$ **4.** $33; 110$

Daily Homework Quiz

1. $2, 4$ **2.** $-\frac{1}{3}, 5$ **3.** $x < -6$ or $x > -1$

4. $-0.8 \le x \le 2$

5. $|x - 32| \le 9$

Lesson Opener

Allow 15 minutes.

1–3. Inequality 1, D; Inequality 2, C; Inequality 3, A; Inequality 4, B

Practice A

1. yes   **2.** no   **3.** no   **4.** yes   **5.** no   **6.** yes

7. no, no   **8.** no, yes   **9.** no, no   **10.** yes, no

11. yes, yes   **12.** yes, yes   **13.** yes, no

14. no, no

15.

16.

17.
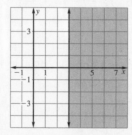

18.
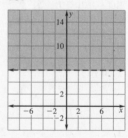

Lesson 6.5 *continued*

19.
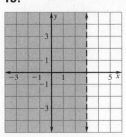

20.
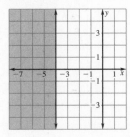

21.
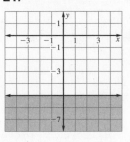

22.
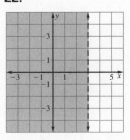

23.
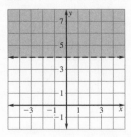

24.
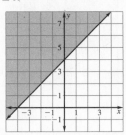

25.
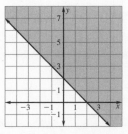

26.
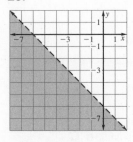

27.

28.
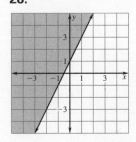

29.
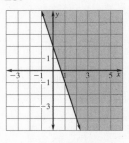

30. $x + y \leq 40$
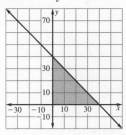

31. $2x + 2y > 16$
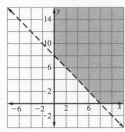

32. $2x + y \geq 26$
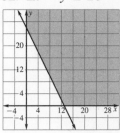

Practice B

1. yes, no **2.** yes, yes **3.** no, yes **4.** yes, yes

5. no, no **6.** yes, no **7.** yes, yes **8.** yes, yes

9. no, yes **10.** no, no

11.
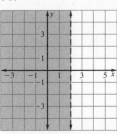

12.
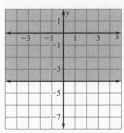

13.
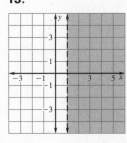

14.
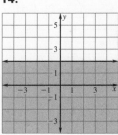

Lesson 6.5 *continued*

15.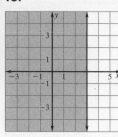

16.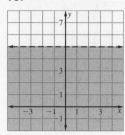

17.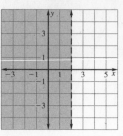

18.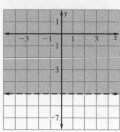

19.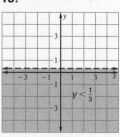

20.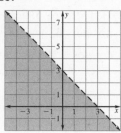

21.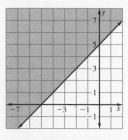

22.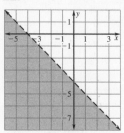

23.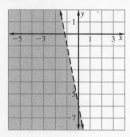

24.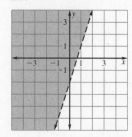

25.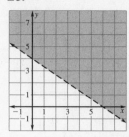

26.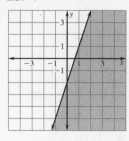

27.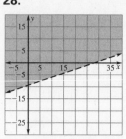

28.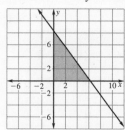

29. $2x + 3y \geq 24$

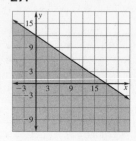

30. $20x + 15y \leq 125$

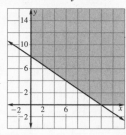

31. $2x + 4y < 28$

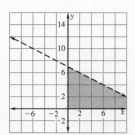

Practice C

1. no, yes **2.** no, no **3.** no, yes **4.** no, yes

5. no, no **6.** yes, no **7.** yes, no **8.** no, yes

9.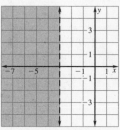

10.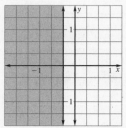

Algebra 1
Chapter 6 Resource Book

Lesson 6.5 *continued*

11.

12.

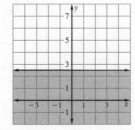

13.

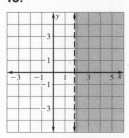

14.

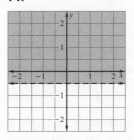

15.

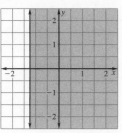

16.

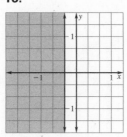

17.

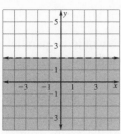

18.

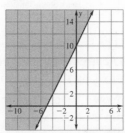

19.

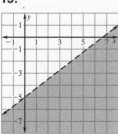

20.

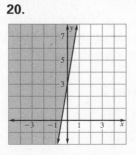

21.

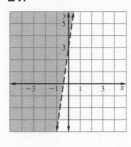

22.

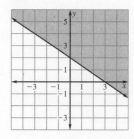

23.

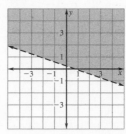

24.

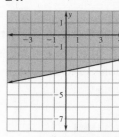

25.

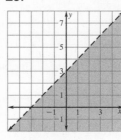

26.

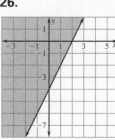

27.

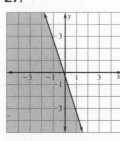

28.

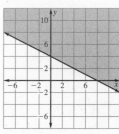

29.

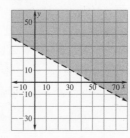

30. $3x + y \le 65$

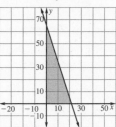

Lesson 6.5 *continued*

31. $20x + 80y \leq 3000$ **32.** $0.35x + 0.8y \leq 12$

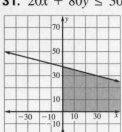   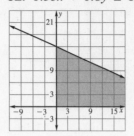

Reteaching with Practice

1. $(0, 0)$ is not a solution; $(-1, -2)$ is a solution.   **2.** $(2, 2)$ is a solution; $(-2, 2)$ is not a solution.

3.
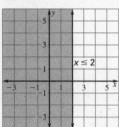

4.
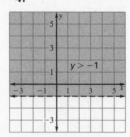

5.
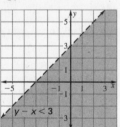

6.
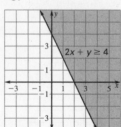

7.
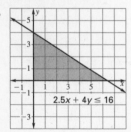

Interdisciplinary Application

1. $3.75x + 4.5y < 1000$

2.
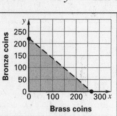

3. $3.75x + 4.5y < 750$ **4.**
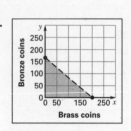

Challenge: Skills and Applications

1. $y \leq \frac{2}{5}x + \frac{31}{5}$ **2.** $y < -\frac{1}{3}x + 9$

3. $x + y \leq 90$

4.
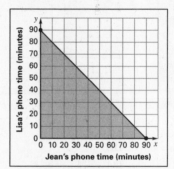

5. The diagonal boundary line would be parallel to the current boundary line, but it would intersect each axis at 120 rather than at 90.

6. Now Lisa can't be on the phone more than 60 min even if Jean uses the phone less than 30 min. The upper left corner of the graph is cut off, leaving a region in the shape of a trapezoid. For $x < 30$, the boundary becomes a horizontal line located at $y = 60$. For $x \geq 30$, the boundary line stays the same.

Quiz 2

1. $7\frac{1}{3}$, -2 **2.** $x \leq -3$ or $x \geq 17$

3. no; $-2 \not> 11$

Lesson 6.5 *continued*

4.

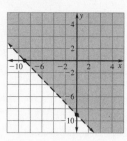

5. $y > \frac{1}{2}x - 2$

Lesson 6.6

Warm-up Exercises

1. 93.25 **2.** 30.25 **3.** 173

Daily Homework Quiz

1. no, yes **2.**

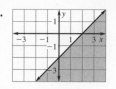

3.

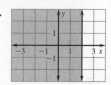

4.

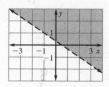

Lesson Opener

Allow 10 minutes.

1. 21, 24, 25, 31, 36, 37, 38, 40, 43, 48, 48, 52, 54, 59, 62

2.

Tens Digit	Data
2	21, 24, 25
3	31, 36, 37, 38
4	40, 43, 48, 48
5	52, 54, 59
6	62

3. *Sample answer:* groups of data; fewer numbers to order **4.** 48; more teachers are 48 than any other age. **5.** 15; 40; seven teachers are older than 40 and seven are younger than 40.

Graphing Calculator Activity

1. a. 127.4 **b.** 2.6 **c.** 2554.4 **d.** 11.6

---

2. The means of parts of (b) and (d) are unreasonable. In part (b), the one person with 9 siblings causes the mean to be higher than the majority of the data. In part (d), the person who has flown 48 times causes the data to be too high.

Practice A

1.
```
2 | 5  6  8
3 | 2  5  6  7
4 | 1  2  4  8
5 | 0  1  8
```
Key: 4|2 = 42

2.
```
0 | 2  3  5
1 | 0  1  4  6  8
2 | 1  2  6  7  9
3 | 3  5
```
Key: 3|3 = 33

3.
```
3 | 2  8  9
4 | 1  2  3
5 | 2  3  3  4  5  8
6 | 1  1  7
```
Key: 6|1 = 61

4.
```
0 | 2  3  5  9
1 | 0  5  6  7
2 | 3  4
3 | 0  1  1  4  8  9
4 | 2
5 | 4  6  9
```
Key: 5|4 = 54

5. mean: 7.9
median: 7.5
mode: 7

6. mean: 9
median: 8
mode: 8

7. mean: 50.9
median: 52
mode: 41

8. mean: 117.7
median: 118
mode: none

9. mean: 112
median: 106
mode: none

10. mean: 151.1
median: 133.5
mode: 111

11. 11, 10, 9, 8, 8, 8, 7, 7, 6, 6, 5, 5, 4, 4

12. mean: 7
median: 7

13. The mean would change to 7.86; the median would still be 7.

14.
```
2 | 8
3 | 0  0  2  5  6  7  7  7
4 | 9
5 | 7
6 | 4  7
```
Key: 6|4 = 64

Lesson 6.6 *continued*

15. mean: 41.5

median: 37

mode: 37

16. The median or mode best represents the data because the data is centered around 37.

Practice B

1.
```
2 | 0 1 1 3
3 | 5 7 9
4 | 1 8
5 | 0 2 3
6 | 4 6 6 8
```
Key: 6|4 = 64

2.
```
0 | 2 2 8
1 | 0 1 5 5 6 9
2 | 0 3 5 7
3 | 1 2 4 4 6
```
Key: 3|1 = 31

3.
```
1 | 0 1 3 5 7 8
2 | 0 2 2 3 4 8 8 9
3 | 0 3 5 6 7 9
```
Key: 3|0 = 30

4.
```
0 | 2 2 5 6 8 8
1 | 0 7 8
2 | 1 4 8 9
3 | 0 2 3 8
4 | 1 3
5 | 2 4 5
```
Key: 5|2 = 52

5. mean: 89

median: 90

mode: 90

6. mean: 37

median: 37

modes: 36, 37, 38

7. mean: 9

median: 9

mode: 8

8. mean: 54.1

median: 54

mode: none

9. mean: 58.$\overline{5}$

median: 59

mode: 60

10. mean: 47

median: 48

mode: 48

11.
```
5 | 7 8 9 9
6 | 1 2 6
7 | 0 1
8 | 1 5 6 6 8
```
Key: 5|7 = 5.7

12. mean: 7.1

median: 6.8

13.
```
2 | 05 69
3 | 55
4 | 14 18 23 35 42
```
Key: 4|14 = 4.14

14. mean: $3.70

median: $4.16

mode: none

15. The median best represents the data because the data is centered around $4.16.

Practice C

1.
```
2 | 1 8
3 | 3 5
4 | 2 5 7 7
5 |
6 | 2 4 4 6 7
7 | 0
8 | 5 9
9 | 1 3 4
```
Key: 9|1 = 91

2.
```
5 | 1 5 5 6 7 9
6 | 0 1 3 4 8 9
7 | 3 3 4 7
8 | 0 3 4
9 | 2 5
```
Key: 9|2 = 92

3.
```
12 | 3 5 7
13 | 3 6 6 7 9
14 | 0 1 4 7
15 | 0 6 9
```
Key: 15|0 = 150

4.
```
0 | 6 9
1 | 5 5 6
2 | 2 2 6 7 9
3 | 1 5 5 7
4 | 1 7 7
```
Key: 4|1 = 4.1

5. mean: 49.8

median: 46

modes: 46, 25

6. mean: 329.4

median: 332.5

mode: 400

7. mean: 1362.3

median: 1398

mode: 1398

8. mean: 967.4

median: 952

mode: 925

9. mean: 66.9

median: 60.8

mode: 60.8

10. mean: 149.6

median: 147.2

mode: none

11. mean: 30.75

median: 29.03

Lesson 6.6 *continued*

12. Data above the median is farther from the median than the data below the median.

13.

13	4.5
14	4.8 5.9 8.0
15	3.6 7.2
16	0.9 2.2 7.6
17	0.7 6.5
18	6.0

Key: $16|2.2 = 162.2$

14. mean: 158.99

median: 159.05

mode: none

15. The median or the mean best represents the data because the data is centered around 159 mph.

Reteaching with Practice

1. 2   8   10   13   16   22
28   31   35   35   50   56

2. Ordered stem-and-leaf plot

	0	2 2 7
Stem	1	1 6 9 Leaves
	2	6 7
	3	3 8 9 Key: $2|6 = 26$

The data in increasing order:
2   2   7   11   16   19
26   27   33   38   39

3. mean = 20, median = 19, mode = 2

Cooperative Learning Activity

1. Answers will vary depending on the surveys conducted.   **2.** Circle graphs report information in an easy-to-read and easy-to-understand way. Line graphs help show how data changes over time. Bar graphs show the comparison between different groups of data. Box-and-whisker plots and stem-and-leaf plots are good for displaying the distribution of data in a set.

Interdisciplinary Application

1–6. Answers will vary.

Challenge: Skills and Applications

1. about 36.6, 36, 34   **2.** the median is unchanged, the mean decreases, and the mode becomes 40   **3.** no; the missing salary cannot be $27,000 or $32,000   **4.** $x \leq 32$ **5.** yes; $41,000   **6.** $0.8 billion, $1.3 billion, $3.0 billion, and $4.4 billion   **7.** $3.95 billion   **8.** about $4.38 billion   **9.** $5.4 billion

Lesson 6.7

Warm-up Exercises

1. 14   **2.** 10   **3.** 13

Daily Homework Quiz

1.

4	5 7 8
5	2 4 5 5 6 7 9
6	1 2 3
7	0

Key: $6|1 = 61$

56, 55.5, 55

2.

11.1	9
11.2	8
11.3	5 7 9
11.4	2 8
11.5	6 6 7 8
11.6	1 2 7 9

Key: $11.5|6 = 11.56$

11.49, 11.56, 11.56

Lesson Opener

Allow 10 minutes.

1. Check work.   **2.** 5, 8, 9, 10, 11, 12, 14, 16, 19, 22, 25   **3.** 12 should remain in the top row; it is the median.   **4.** 9 should remain in the second row; it is the median of the numbers in the first section.   **5.** 19 should remain in the second row; it is the median of the numbers in the second section.   **6.** 4; there are same number of cards, 2, in each section.

Practice A

1. b   **2.** c   **3.** d   **4.** a   **5.** 7.5, 10, 13.5

6. 74, 79, 88   **7.** 23, 27.5, 30   **8.** 49, 57, 61

9.

Lesson 6.7 *continued*

10.

10 20 30 40 50 60 70 80
15   41 48   67 75

11.

500 600 700 800 900 1000
561 610 698 800   979

12.

```
4 | 2 3 6 6 7 8 9 9
5 | 0 0 1 1 1 1 2 2 4 4 4 4 4 5 5 5 5
  | 6 6 6 7 7 7 7 8
6 | 0 1 1 1 2 4 4 5 8 9
```
Key: 4|2 = 42

13. 51, 55, 58

14.

40 45 50 55 60 65 70
42   51 55 58   69

15. Answers may vary.   **16.** 3 hours

17. Yes; both represent 25% of the adults.

Practice B

1. c   **2.** d   **3.** a   **4.** b

5.

0 8 16 24 32 40 48
0   15   28   38 46

6.

10 20 30 40 50 60
11   24 32   42   55

7.

50 150 250 350 450 550
68 77 92 116.5   591

8. Answers may vary.   **9.** Answers may vary.

10.

```
4 | 2 3 6 6 7 8 9 9
5 | 0 0 1 1 1 1 2 2 4 4 4 4 4 5 5 5 5
  | 6 6 6 7 7 7 7 8
6 | 0 1 1 1 2 4 4 5 8 9
```
Key: 4|2 = 42

11. 51, 55, 58

12.

40 45 50 55 60 65 70
42   51 55 58   69

13. Answers may vary.

14. A new data point of 35 would make the left whisker longer.

15. 6 hours

16. Yes; both represent 25% of the students.

Practice C

1. b   **2.** c   **3.** d   **4.** a

5.

30 40 50 60 70 80 90 100 110
36 46   71   85   105

6.

2   4   6   8   10 12 14
2.0 3.9 4.8 7.7   12.7

7.

90 120 150 180 210 240 270
97.1   116.1   240.8
100.8 107.45

8. Answers may vary.   **9.** Answers may vary.

10.

```
3 | 1 1 3 6 7
4 | 2 5 5
5 | 1 1 4
6 | 0
```
Key: 6|0 = 60

11.

```
2 | 5 5 9
3 | 3 4 5 9
4 | 1 2 5 9
5 |
6 |
7 |
8 | 0
```
Key: 8|0 = 80

12.

30 35 40 45 50 55 60
31 34.5 43.5 51   60

13.

20 30 40 50 60 70 80
25 31 37 43.5   80

14. Answers may vary.   **15.** You cannot tell what the mean is from a box-and-whisker plot.

16. false

Lesson 6.7 continued

Reteaching with Practice

1.

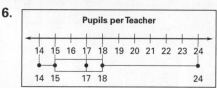

A number line from 0 to 25, with marks at 1 (Least number), 8 (First quartile), 13 (Second quartile), 23 (Third quartile), 25 (Greatest number).

2.
about 92 million

Real-Life Application

1.

7	5 9
8	1 2 5 8 9
9	4 5 6 6 9

Group 1

7	3 6 8 9
8	1 3 5 6
9	0 1 3 6

Group 2

6	9 9
7	0 2 3 5 6
8	0 1 5 7
9	1

Group 3

2.

Box-and-whisker plot (70–100): 75, 81.5, 88.5, 95.5, 99

Box-and-whisker plot (70–100): 73, 78.5, 84, 90.5, 96

Box-and-whisker plot (70–100): 69, 71, 75.5, 83, 91

3. Group 1 = Mean 88.25; Median 88.5;
Group 2 = Mean 84.25; Median = 84;
Group 3 = Mean 77.3; Median = 75.5; Yes

Challenge: Skills and Applications

1.

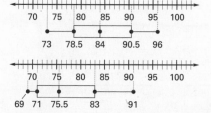

Stadium Capacity (thousands): box-and-whisker plot from 60 to 80, with 60, 64, 70, 75, 80.

---

2. *Sample answer:* The data are pretty evenly distributed. The quartiles are all about the same size. **3.** 60 thousand, 66 thousand, and 73 thousand; stem-and-leaf plot; data are lost in the box-and-whisker plot. **4.** 70,000; box-and-whisker plot; the median is marked in the box-and-whisker plot.

5. about 69, 533; stem-and-leaf plot; data are lost in the box-and-whisker plot.

6.

Pupils per Teacher: box-and-whisker plot from 14 to 24, with 14, 15, 17, 18, 24.

7. *Sample conclusions:* Three-quarters of the states are clustered within a fairly narrow range, with 14–18 pupils per teacher. The one-quarter of the states with the most pupils per teacher shows a greater spread than the entire other three-quarters of the states, with anywhere from 18 to 24 pupils per teacher.

Review and Assessment

Review Games and Activity

1. $x > -8$ **2.** $x \geq -9$ **3.** $x \leq -3$
4. $x \leq -2$ **5.** $x < -12$ or $x > 2$
6. $0 \leq x \leq 16$ **7.** $x > 7$ **8.** $-6 \leq x < 7$
9. all real numbers   WIND SOCKS

Test A

1.
Number line with closed dot at −3.

2.
Number line with open dot at 3.5.

3. $x \leq 4$
Number line with closed dot at 4.

4. $a < 7$
Number line with open dot at 7.

5. $x > 45.5$ **6.** $x \leq 5$ **7.** $x < -3$
8. $12.99x \leq 45$
9. $1 < x < 4$
Number line with open dots at 1 and 4.

10. $2 < x \leq 7$
Number line with open dot at 2 and closed dot at 7.

11. $x < -2$ or $x > 3$
Number line with open dots at −2 and 3.

12. $-1 < x < 4$

Review and Assessment *continued*

13. $x \geq 5$ or $x \leq 1$ **14.** $5, -5$

15. $-7 < x < 3$ **16.** yes   **17.** no

18.

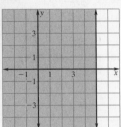

19.

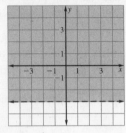

20.

	2	0 2 7	
Stem	3	0 3 6	Leaves
	4	0 2	
	5	4	Key: $4|0 = 40$

21. Mean: 6; Median: 6; Mode: 9

22. First quartile: 1; Second quartile: 4; Third quartile: 8

Test B

1. $p \geq -2$

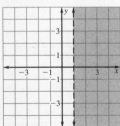

2. $y > -4$

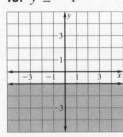

3. $x < \dfrac{1}{18.25}$ or $x < 0.055$ **4.** $x \geq \dfrac{1}{4}$ **5.** $x \geq -5$

6. $1.50x \leq 200$

7. $x > 1$ or $x < -2$

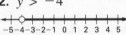

8. $-4 \leq x < 3$

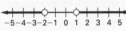

9. $x < -\dfrac{1}{2}$ or $x \geq 1$

10. $-1 < x \leq 3$ **11.** $x > 0$ or $x < -3$

12. $1, -7$ **13.** $8, -4$ **14.** $2 < x < 8$

15. $x \geq 7$ or $x \leq -10$ **16.** No   **17.** Yes

18. $x > 1$

19. $y \leq -1$

---

20. $0.99x + 0.49y \leq 5$

21.

	2	1 5 6	
	3	1 4 7	
Stem	4	5	Leaves
	5	1 4	Key: $2|1 = 21$

22. Mean: 8
Median: 8
Mode: 8

23. First quartile: 4
Second quartile: 7
Third quartile: 12

24.

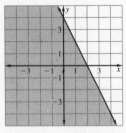

Test C

1. $x \leq -15$

2. $b \geq -4$

3. $x < \dfrac{1}{5.05}$ or $x < 0.198$ **4.** $x > 12$ **5.** $x \leq -2$

6. $1.50 + 0.70x \leq 3.50$

7. $x \geq 5$ or $x < 0$

8. $-\dfrac{9}{2} < x \leq \dfrac{7}{2}$

9. $x \leq -2$ or $x > 3$

10. $0 \leq x < 4$

11. $x > 3$ or $x \leq 0$ **12.** $7, -\dfrac{31}{3}$

13. $x \leq -3$ or $x \geq 10$ **14.** Yes   **15.** No

16. $y \leq 4 - 2x$

17. $y > 3 - 2x$

18. $50x + 20y \leq 1000$

19.

	2	2 3 3	
	4	1 2 5	
Stem	5	5 7	Leaves
	6	1	Key: $4|1 = 41$

20. Mean: 3.1
Median: 3.2
Mode: None

21. First quartile: 2.5
Second quartile: 3.9
Third quartile: 5.3

Review and Assessment *continued*

22.

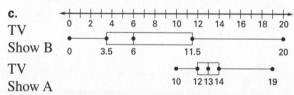

18 20 22 24 26 28 30 32 34 36
19 21   25   34 36

SAT/ACT Chapter Test

1. D **2.** C **3.** D **4.** A **5.** B **6.** C **7.** B
8. A **9.** C

Alternative Assessment

1. Complete answer should address these points:
a. In solving a linear inequality, the inequality symbol is reversed when dividing or multiplying both sides by a negative. **b.** For a linear inequality in two variables, the line is dashed for $<$ or $>$ and solid for $\leq$ or $\geq$. Also, one has to determine which way to shade. The solution of an inequality is a shaded region, whereas the solution of an equation is just a line.

2. a.

0	0 0 1 2 3 3 4 4 4 5 5
	6 6 6 6 7
1	0 0 1 2 4 5 5 9
2	0 Key 1 | 5 = 15

b. 7.52; 6; 6; median and mode; *sample explanation:* Most of the data seems to be clustered around 6.

c.

TV
Show B

0 2 4 6 8 10 12 14 16 18 20
0   3.5   6   11.5   20

TV
Show A

10 12 13 14   19

d. *Sample answer:* Less than 25% of TV Show B viewers are more than 12 years old, whereas 75% of TV Show A viewers are more than 12 years old. **e.** $0 \leq x \leq 20$ **f.** no; 75% of the viewers are between 10 and 14 years old. **g.** *Sample answer:* The ages of viewers of TV Show B are spread out, whereas the audience for TV Show A is more tightly grouped in the 12 to 14 year old range. TV Show A would be better for a product that is geared towards a specific age group, whereas TV Show B would be fine if advertisements are geared toward a wider age group. The median age of TV Show A viewers is 13 years, but only 6 years for TV Show B.

3. Yes, it's close to the median; 14.58 years; no; *sample explanation:* 45 years is not representative of the data and affects the mean by too much.

Project: Not What They Used to Be

1. Check that x is between the lengths of the two pre-1970 movies inclusively. *Sample answer:* $174 \leq x \leq 239$ (minutes) **2.** Check computations. *Sample answer:* 120 min; about 121 min; about 165 min **3.** Check students' number lines; *Sample answer:* none of the means are solutions to the inequality. **4.** Check students' box-and-whisker plots; *Sample answer:* one pre-1970 movie is in the upper quartile but the other is not.

Sample answers are based on the following data.

Gone With the Wind (1939) 239 min

The Sound of Music (1965) 174 min

The Exorcist (1973) 120 min

Jaws (1975) 119 min

Star Wars (1977) 121 min

Raiders of the Lost Ark (1981) 115 min

ET: The Extra-Terrestrial (1982) 115 min

Return of the Jedi (1983) 132 min

Jurassic Park (1993) 127 min

Forrest Gump (1994) 179 min

Titanic (1997) 188 min

Information came for the Library of Congress catalog Web site and from http://www.cinemedia.net.

Cumulative Review

1. yes **2.** yes **3.** no **4.** yes **5.** no
6. yes **7.** $-15, -2.9, -\frac{100}{35}, \frac{13}{30}, 8, 10$
8. $-5.6, -5.069, -5.05, -5.003, -0.506, 5.5$
9. $-3\frac{6}{7}, -3.6, -3.023, -3, 3.25, 3\frac{2}{5}$
10. $-9.01, -8.9, -\frac{34}{9}, -1, -\frac{4}{17}, 9$
11. $\begin{bmatrix} 0.92 & -10.29 \\ 0.14 & 0.91 \end{bmatrix}$ **12.** $\begin{bmatrix} -10\frac{1}{3} & \frac{8}{9} & -2 \end{bmatrix}$

13. $\begin{bmatrix} \frac{29}{30} \\ -\frac{5}{6} \\ \frac{5}{2} \end{bmatrix}$ **14.** $\begin{bmatrix} 0.5 \\ -0.8 \end{bmatrix}$ **15.** 129 **16.** $\frac{33}{27}$

17. $6\frac{1}{10}$ **18.** 0.088 **19.** -3 **20.** -1.84
21. $-\frac{6}{5}$ **22.** 0 **23.** $\frac{10}{3}$ **24.** 5

Answers

25. x-intercept $= 0$, y-intercept $= 0$

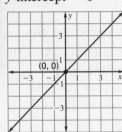

26. x-intercept $= 4$, y-intercept $= -12$

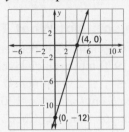

27. x-intercept $= \frac{1}{3}$, y-intercept $= -3$

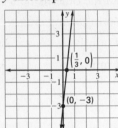

28. x-intercept $= 13$, y-intercept $= -\frac{13}{6}$

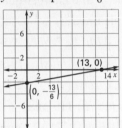

29. $y = -\frac{5}{2}$, $m = 0$, $b = -\frac{5}{2}$

30. $y = 0.271x - 1.875$, $m = 0.271$, $b = -1.875$

31. $y = -\frac{4}{3}x + 4$, $m = -\frac{4}{3}$, $b = 4$

32. $y = x - 9$, $m = 1$, $b = -9$

33. $y = \frac{1}{3}x - 7$, $m = \frac{1}{3}$, $b = -7$

34. $y = 0.25x + 0.25$, $m = 0.25$, $b = 0.25$

35. $y = 9x - 45$ **36.** $y = x + 5$

37. $y = -25x - 26$ **38.** $y = -\frac{11}{2}x - \frac{111}{2}$

39. $y = -\frac{2}{3}x$ **40.** $y = -\frac{2}{3}x + 1$

41. $y = -\frac{5}{6}x + \frac{5}{2}$ **42.** $y = 4x - \frac{1}{2}$

43. $y = \frac{5}{3}x - \frac{19}{6}$ **44.** $y = -x + 6.4$

45. $9x + y = -81$ **46.** $27x - 20y = -285$

47. $x + 2y = 10$ **48.** $18x + 5y = 21$

49. $6x + y = -21$ **50.** $8x - 2y = -21$

51. $t > -8$ **52.** $x \leq 35$ **53.** $y < -1.759$

54. $x \geq 768$ **55.** $s > 22$ **56.** $-6.5 \leq m$

57. $-1 \leq x \leq 9$

58. $x < \frac{1}{2}$ or $x > 12$

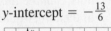

59. $x < -2$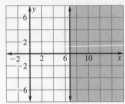

60. $-\frac{1}{3} \leq x < \frac{2}{3}$

61. $x < -11$ or $x > 8$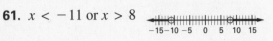

62. $x \leq -6$ or $x \geq 3$

63. $-5, 11$ **64.** $-2, 2$ **65.** $\frac{11}{30}, \frac{13}{30}$

66. $16.3, -3.3$

67. $x \geq 6\frac{3}{4}$ **68.** $y > -5$

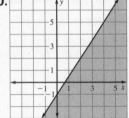

69.

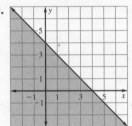

70.

71.

1	2
2	2 5 6
3	6
4	
5	
6	
7	8
8	9

12, 22, 25, 26, 36, 78, 89
Key: $3|6 = 36$

72.

1	0 1 2 4 9
2	3

10, 11, 12, 14, 19, 23
Key: $2|3 = 23$

73.

5	4 6
6	
7	2 3
8	7 9

54, 56, 72, 73, 87, 89
Key: $7|2 = 72$

74.

10	0
11	2
12	5
13	
14	8
15	2
16	8

100, 112, 125, 148, 152, 168
Key: $14|8 = 148$